PLUMBING
A HOUSE

PLUMBING A HOUSE

PETER HEMP

The Taunton Press

The Taunton Press
Inspiration for hands-on living™

10 9 8 7 6 5
Printed in the United States of America

The Taunton Press, 63 South Main Street, PO Box 5506,
Newtown, CT 06470-5506
e-mail: tp@taunton.com

For Pros / By Pros®: Plumbing a House was originally
published in 1994 by The Taunton Press, Inc.

For Pros / By Pros® is a trademark of The Taunton Press, Inc.,
registered in the U.S. Patent and Trademark Office.

Library of Congress Cataloging-in-Publication Data

Hemp, Peter A. (Peter Addison)
    For Pros / By Pros®: Plumbing a house / Peter Hemp.
        p.   cm.
    Includes index.
    ISBN 1-56158-333-2
    1. Plumbing — Amateurs' manuals. 2. Plumbing. I. Title.
TH6124.H447   1998   94-33843
696'.1—dc20   CIP

ACKNOWLEDGMENTS

The energies of many good folks have gone into this book, and it is impossible to acknowledge everyone who has been helpful and interested in this project. However, I would like to single out two companies for special mention; without them this book would never have been finished. The Rubenstein Supply Company of Oakland, California, gave me a free run of their facilities and inventory and researched many products and suppliers for me. Sane Contracting of Albany, California, allowed me access to their job sites and even held up their production schedules so that I could get photographs. I'd also like to thank Meyer Plumbing Supply and Moran Supply, also of Oakland, and Ohmega Salvage, The Sink Factory, and Urban Ore of Berkeley, California, for their great generosity with their time, space, materials and photographic opportunities. Another debt that I cannot repay is to the Homemade Cafe of Berkeley and the Baywolf Restaurant of Oakland for providing me with field offices, gratis, for the duration of this project.

The following companies and organizations were very helpful in aiding me with technical and product information: American Brass and Iron of Oakland and Los Angeles; Colonial Engineering, Kalamazoo, Michigan; The Copper Development Association, New York City; Ellis Hardware, Oakland; Eslon Thermoplastics, Charlotte, North Carolina; G & G Hardware, Oakland; H & H Mining Co., Ocean Beach, California; The Hammer and Chisel, Berkeley; Harvel Plastics, Easton, Pennsylvania; International Association of Plumbing and Mechanical Officials, Walnut, California; J.W. Landscape Co., Berkeley; Maskell-Robbins, Watsonville, California; Phillips Driscopipe, Richardson, Texas; Plastic Pipe and Fittings Association, Glen Ellyn, Illinois; Plastics Pipe Institute, New York City; R.G. Sloan, Little Rock, Arkansas; The Society of the Plastics Industry, Wayne, New Jersey; State of California Department of Health Services, Berkeley; and Valley Products Ltd., Nottingham, England. Information presented in the charts in chapters 3 and 7 has been adapted from the Uniform Plumbing Code.

And of course, not least but last, the following individuals availed themselves in many ways, helping to make this book a reality. I appreciate all of you. Thank you very, very much.

Donald E. Anderson, Jeff Beneke, Norm Berzon, Ed Botts, Steve Brewer, Miriam Camp, Everette Campbell, Jon Carroll, Gerald Casey, Martha Casselman, Ray Chandler, Danny Cline, Karen M. Connelly, Bill Dane, Ruth Dobsevage, Conrad Dusseldorf, Larry Goldman, Daniel Hemp, Katherine R. Hemp, Mary E. Hemp, Janet Hinze, Susan Hoffman, John Halliwill, Jimmy Kane, Elsie Krem, Ron Kyle, Steve Mastin, Chuck Miller (can we go fishing now, Chuck?), Laura Mooney, Neil Moran, Kenneth Neuberger, John Newell, Joanne Okamoto, Elizabeth Overmyer, Roger Patzer, David Peoples, Janet Peoples, Pat Pikul, E. Rene, Marty Reutinger, Virginia Robles, Bob Rubenstein, Craig Rubenstein, Gardenia Sarazin, Tucker Seymour, John Sloan, Pete Smith, Denny Smithson, Jane Spangler, Allan Temko, Becky Temko, Sandra Tillin, John Traylor, Gene Turitz, Chris Utter, Don Villa, Dick Wagner, Jimmy Weyhmiller, Owen Whetzel and Michael Wild.

And thanks again to my good buddy Leo L.

INTRODUCTION

I'm a plumber based in Albany, California (a town across the bay from San Francisco), and this is my twenty-second year in the plumbing trade. Some years ago I started writing magazine articles about plumbing. I've also written an elementary plumbing-repair manual for home owners. The book you now hold in your hands is part of my latest attempt to demystify the design and installation of residential plumbing systems. This book, the first of a two-volume set, covers rough plumbing — the installation of piping systems that serve the plumbing fixtures and plumbing-related appliances in the home — as well as remodeling. The companion volume, *Installing and Repairing Plumbing Fixtures,* discusses finish plumbing — the installation of fixtures and appliances and their out-of-wall support plumbing — as well as troubleshooting problems such as clogs and leaks.

This book covers the four basic plumbing systems: drain, waste and vent piping (DWV); fresh-water service, supply and distribution piping; fuel-gas supply piping; and gas-appliance venting. It also discusses tools needed for plumbing, types of pipes and fittings and their applications, pipe joinery and repair and remodel plumbing. The book concludes with a list of suppliers and manufacturers mentioned in the text.

Plumbing systems must be designed and installed in accordance with national and local standards, called codes, and your work will be checked by the local plumbing inspector when all the piping systems are in place. The inspection involves a water or air test of the DWV and fresh-water systems and an air test on any gas piping, as well as a close look at piping materials, pipe sizes and joints, all of which must conform to the code. Any deficiencies must be remedied and reinspected; no further construction can take place until the inspector signs off on the plumbing. A second purely visual inspection is done later, after all the fixtures and traps have been installed.

Plumbing codes are a real maze. There are three major national codes I am familiar with: the Uniform Plumbing Code (UPC), the BOCA National Plumbing Code and the Standard Plumbing Code (SPC), which is sometimes called the Southern Code. Additionally, every community has its own local code. What is happily accepted in one area may be prohibited in others. What one very competent plumber has done quite successfully over the years might be frowned upon by an inspector in another part of the country. Officials are aware of the problem, and are working toward reconciling code requirements among the three major codes. Perhaps someday we'll have a national plumbing code, and books like this one will be easier to write and interpret.

Codes are minimum standards, intended to guarantee a certain level of workmanship. They are not always the best standards. Just because your house has been plumbed to code does not mean you will be happy with the results. Even though your water pipes may be sized large enough to satisfy both your major and local code as well as the local inspector, you still may not get a comfortable shower on the second floor. Drainage pipes may be sized and vented to code, but they may still be annoyingly noisy in certain locations.

I live and work in a high-density population area served by a municipal water supply, so I do not deal with artesian wells or water pumps, which are common in more rural parts of the country. Since the entire area where I work is served by municipal sewers, I have no call to install septic systems either. When I worked in rural areas, I did install drainage piping to the building sewer and make the connection between the building sewer and the newly installed septic tank, but I did not excavate for or set the septic tank itself. That was years ago, and because I do not consider myself an authority on these aspects of plumbing, I have not discussed them in this book.

Rough plumbing isn't very often installed by non-plumbers other than an occasional general contractor. Non-plumbers are more apt to attempt finish plumbing. What I am trying to accomplish with these books is to give interested non-plumbers the information they need to do some or all of the job themselves and to give builders a means to judge the work of the plumbers that they hire.

If you are a builder or tradesperson who is considering plumbing a new home or doing a major remodel, it pays to familiarize yourself with the major code for your area. (Plumbing codes can usually be checked out of public libraries.) Ask your building inspector about local codes that supersede the major code. These usually take the form of short photocopied documents available free or for a small fee. After this book, it could be the best money you spend.

If you are going to do a large plumbing project, ask your local inspector for a job-site consultation before you begin work, so you can resolve any questions you might have concerning your project-to-be. Have your set of approved plans on hand when the inspector arrives so the two of you can review them. If this is your first major plumbing project, it wouldn't hurt to say so. Record the inspector's answers directly on the plan.

Sometimes an odd situation seems to call for an unusual plumbing solution — perhaps the use of a new material or an uncommon arrangement of pipe and fittings. If you aren't sure whether a particular material or method of construction is acceptable, consult your local inspector. My major code (the UPC) gives inspectors the authority to approve alternative materials and methods of construction, provided that the inspector thinks they are the equivalent of code-sanctioned procedures in quality, safety, strength, effectiveness and durability.

Once a house is completed, you do not see the rough plumbing, but if it was not installed properly you'll be reminded of that fact every day. Some of the hallmarks of a high-quality rough-plumbing job are an efficient, silent and durable DWV system; a water service, supply and distribution system that brings copious amounts of fresh water at desired temperatures to every fixture; an abuse-resistant fuel-gas supply system that is properly sized and installed; and a water-heater vent system sized to accommodate present and future needs.

Good plumbing is expensive, both in labor and in supplies, and spec-house builders are rarely interested in top-of-the-line work. They view rough plumbing not unlike linoleum: spend just enough to sell the job. If you are more than $200 over the next bid, you lose. But for those who are building a home on which to maintain a solid reputation, or even better, a home to live in, bid is not often the best way to go. The greatest enemies of high-quality work are low bid and haste, usually in that order. A plumbing system contracted for on a time-and-materials basis will sometimes cost more, but often not that much more, and it is money well spent.

The methods discussed in this book have worked satisfactorily for me, year in and year out, so they should also work for you. However, they are not the only way to install a good plumbing system. I have tried to discuss options for materials and methods that I do not employ myself, since what I do might not prove as satisfactory in your area due to different weather, soil and water-quality conditions. I'm not a forecaster of doom for those who find my methods wanting. If I have done nothing more than make you think and induce you to perform your own investigations (the basis of all learning), then I have done you a good turn.

One final word. Remember that every job has a beginning and an end. The end may seem like a long time in coming, especially if plumbing is a new experience for you. If you have the stamina to do the whole job, great. But even if you do only some of the work, I think you'll find that the information in this book will serve you well.

TO MY WIFE AND FAMILY, WHO HAVE SACRIFICED SO MUCH FOR THIS BOOK

CONTENTS

TOOLS

Plumbing a house requires a fair amount of equipment. You may already own many general hand tools, devices for measuring and marking and some safety gear if you do any carpentry or auto and home repairs. But serious plumbing also requires a lot of specialized tools in various models and sizes, including tools for gripping, cutting, drilling and joining. As a plumber with more than 20 years experience in the trade, I have my favorite makes and models; you'll find manufacturers' addresses for the tools mentioned in this chapter listed in Sources of Supply, which begins on p. 206.

GENERAL HAND TOOLS

In rough and remodel work, a hammer is indispensable, and not only for pulling nails. I often use mine as a digging tool, for clearing away earth from underneath soil and sewer pipes, for splitting wood between saw cuts and for knocking out blocking between joists. I find that the hammer with the straightest claw is the handiest — hammers with claws that curve sharply downward do a poor job at

these tasks. A hammer is also useful for pounding out severed pieces of iron and terra-cotta sewer pipe after the snap cutter has done its job.

Wooden hammer handles, though easier on their user, do not stand up to heavy-duty use as well as fiberglass handles. I like my Ridgid model #216 16-oz. bright-red ripping-claw fiberglass-handled hammer because of its rugged construction, and it's easy to spot amidst the piles of rubble that I create with it. (A 20-oz. model is also available.)

Cat's paws frequently come in handy for tasks like removing nails that are in the way of plumbing runs. You can get by with a short-handled cat's paw, if it is a good one like the Estwing HC-10, but it helps to have a longer-handled one, too, for use in areas where the added length is not a hindrance. The Estwing HC-10 has sharply pointed claws that drive well into the wood. The angle and the narrowing gap in the claw has been machined with precision, so the tool makes a great nail gripper. The pounding surface is well placed for the driving hammer.

Crowbars and pry bars are very useful, especially in remodel plumbing. I have a 4-ft. long, standard crowbar for pulling spikes, prying apart wood construction and lifting bathtubs for removal. I also have a smaller pry bar, the American-made OTC #7168, which I use for moving tubs once I have lifted them with the big crowbar; I also use it for digging and for removing blocking, especially between floor joists and at rim joists. I keep the chisel point of this bar well honed so I can throw it well into a block and then bust it out with a little leverage.

Wood chisels are another frequently used tool. In rough and remodel plumbing, the 1½-in. chisel gets most of the use removing wood between saw cuts made vertically on joists, especially rim joists on exterior plumbing walls. Wood chisels should have a full-length tang with a metal impact surface on the end of the handle.

I also carry several sizes of cold chisels for use on rough and remodel jobs. The handiest are the ⅜-in. and the ¾-in. chisels. The ¼-in. chisel is useful for cutting through any nails that are encountered when drilling holes in stud walls.

When blasting away exterior stucco to make a hole for a heat-vent thimble, I use ¾-in. and 1-in. chisels. A dull chisel is a waste of time; it pays to have several chisels in each size so you can have one or two in your box while the others are at the sharpening service with your sawblades.

A four-way screwdriver with interchangeable bits is a handy addition to the plumber's tool kit. Of the many available brands, the Pasco #4208 is my favorite because its removable bit shaft and barrel are heavily chrome plated and resist rusting. Also, the Pasco bits are harder and longer, and the squared handle gives better purchase. If you pull the bit shaft out, the driver becomes a nut driver for 5/16-in. no-hub, Mission, Fernco, Calder and similar couplings (see pp. 90-93).

Two other tools for dealing with 5/16-in. fittings are the 5/16-in. nut driver and the 5/16-in. open-end box wrench. The standard 5/16-in. nut driver fits the screw clamps and aids in undoing and assembling all quality band seals. It's a basic but valuable tool. The 5/16-in. open-end box wrench is a very handy rough and remodel tool, a must for loosening and retightening new and existing installations of no-hub couplings and most band-seal couplings (Mission, Fernco, Indiana Seal) that get framed in walls and other hard-to-get-at areas. Diamond and Proto box wrenches are two good brands.

In rough plumbing you have to dig trenches outside the house for sewer pipes. The best tool for this job is a shovel with a narrow, tapering blade and a standard-length handle. Since pipes do not require a trench as wide as a foundation trench, this tool saves a lot of time and sweat when trenching for pipe burial. I purchased my present one, an Ames brand #15-580, at a sand and gravel outlet. I also have a little shovel whose small blade locks straight out from or perpendicular to the handle. It's an original Korean War vintage G.I. model with a contoured wooden handle, and I wouldn't trade it for a dozen straight-handled "new issue" versions. With this shovel I chop vegetation, lift dirt out of excavations, and clear earth from underneath soil pipes so the chain from the soil-pipe cutter can be pulled through and wrapped into place.

If your work involves a structure served by a municipal water service, you will need a water-meter key to be able to turn off the water in the event of an emergency. This key, shown in the photo below, is a simple T-shaped tool about 2 ft. long with a slot in the end that fits over a small rectangular domino on top of the main water valve. Buy the stoutest key you can. Usually the cheaper ones are made from small-diameter stock, and some meter valves are so hard to turn on and off that you can twist the cross-handle right off a cheap key.

Some utilities prefer that they be the party to interrupt water service to a structure, probably because their equipment can be damaged by a novice who

A water-meter key shuts off the water service to a house with city water at the main water valve.

tries to perform this task without any direction. However, in a large city or large suburban area, it might take a long time to get the utility company to act, and if a burst pipe is flooding a customer's house, protocol is a non-issue. My utility has published a handbill that tells the customer how to turn off the water service in an emergency.

TOOLS FOR LAYOUT, MEASURING AND MARKING

If you do any carpentry or woodworking, you probably already have the equipment you will need for layout, measuring and marking. If not, the tools mentioned below can be obtained at any building-supply store or by mail order.

The combination square is used in rough and re-model plumbing for laying out holes in framing (especially floor joists and ceiling joists) prior to boring and sawing for the roughing in of water lines, drainage lines and vents.

Levels are used in both rough and remodel plumbing, but I think that I use mine most on below-slab drainage pipe runs. My plumber's torpedo level, a Pasco #4470, has a flat, magnetized strip on one side for attaching to ferrous pipe, and an opposite, grooved side for centering on any rounded surface. With this level I check vertical cast-iron soil, drain and vent lines for plumb, and I check the slope (fall) on all horizontal drainage lines. I also use it to check for plumb on heating-vent pipe while I'm up on a ladder working my way to the soffit. The magnetic side of the level holds itself to the vent pipe while I screw the legs of the pipe strap to the side of the building. I also have a 2-ft. level, which I use for leveling water heaters and for checking pipe slope.

A plumb bob is handy for laying out toilet- and shower-drain drops in slab construction, for aligning bore holes in upper and lower plates, and for locating the passageway in soffits for bringing through heater-vent piping.

For measuring, I use a tape measure and a folding rule. For measuring long lengths of pipe, a 1-in. wide tape is better than a ¾-in. wide tape. For measuring close distances between copper fittings, a folding rule is indispensable. Once you have gotten used to one, you'll never go back to the tape for this application. The most common folding rules are 6 ft. long when

totally extended, and you need to oil the joints occasionally to prevent breakage. Traditionally folding rules are either natural wood or painted white. I prefer the white rule because it is easier to read in dim crawl spaces and dark attics.

A chalk line is handy for marking stud walls that have to be bored for long sections of water-supply, drainage and vent piping. In a pinch, it can also be used as a plumb bob. For chalk lines I have no brand preference. I'm using the same one (an inexpensive one at that) that I purchased in 1968. I also manage to get by with just one color of chalk: red.

In remodel work I often use a pencil compass for scribing holes in walls and on subfloors for pipes and fittings. I use a simple, inexpensive student's model to lay out the closet-flange hole and a bigger home-made version to lay out the hole for heating-vent thimbles and large-diameter pipe drops that have to be cut with a saw rather than drilled.

TOOLS FOR GRIPPING

Plumbers do a lot of gripping of pipes, couplings and fasteners. General things to look for in a good gripping tool are weight, close machine tolerances and size. The tool should fit your hand well and be easy to manipulate — remember that a lot of your work will be in tight quarters, under less than ideal lighting conditions.

PIPE WRENCHES

Pipe wrenches are essential tools for rough and re-model work, and I usually carry five sizes (6 in., 12 in., 14 in., 18 in. and 24 in.) with me. You may think that a 6-in. pipe wrench is something of a joke. Not so. A good-quality 6-in. wrench can grasp a ¾-in. iron pipe. It is small enough to get into cramped spaces to break loose or snug up threaded connections; as the years go by I find more applications for this small but useful tool.

The 12-in. and 18-in. wrenches are good, general-purpose sizes for installing new gas lines and for "uncorking" 1½-in. and 2-in. threaded iron vent lines in remodel work. The 24-in. wrench I use mostly for unthreading stubborn cleanout plugs and unthreading old, stubborn hot-water supply piping at boilers and water heaters. A word of warning: If you use a wrench that's larger than you really need to tighten a threaded pipe joint, you may inadvertently overtighten the pipe or the fitting. Overtightening can result in a stretched or cracked fitting, and the joint will leak.

Since most of my threaded pipe work is small-diameter (½-in. and ¾-in.) gas lines, I use aluminum-bodied pipe wrenches, which don't cause as much arm fatigue as the standard malleable iron versions. If you are planning to do a lot of plumbing work, you might consider getting at least one aluminum pipe wrench. The Ridgid #814 Aluminum HD at the 14-in. length (the most versatile size) is a good choice.

SLIDE-JAW PLIERS

In rough and remodel plumbing, I use three sizes of slide-jaw pliers: 6 in., 10 in. and 12 in. There are definite differences of shape and design among the various manufacturers, and a mix of brands best accomplishes various tasks.

I find the 6-in. size handy for adjusting copper fittings prior to and during the soldering process. The 10-in. slide-jaw pliers is the most frequently used tool in my collection. The Pasco and Craftsman brand 10-in. pliers can grasp and hold all pipe and fitting sizes up to 1½ in. In rough plumbing, the most popular application of these pliers is the initial assembly of ½-in. and ¾-in. dia. threaded pipe and fittings. I have tried every major brand of tool for this task, and the two brands I mentioned perform the best.

I am very disappointed in the Ridgid Company's 10-in. slide-jaw pliers. Its jaws are undersized for standard pipe and fitting sizes, and the slide action in the adjustment process is very cumbersome. At every setting, the handles are uncomfortable to squeeze. Since a great amount of a plumber's work is dealing with drain and waste components, it is very important to have a well-designed tool to work with.

The 12-in. slide-jaw pliers I use mostly for assembling 1½-in. and 2-in. threaded pipe and fittings, prior to a final snugging with pipe wrenches. For this work the 12-in. Ridgid slide-jaw pliers is fine.

Pipe wrenches are used to snug up pipe joints and to disassemble threaded pipes and fittings.

Slide-jaw pliers are handy for assembling threaded pipe joints.

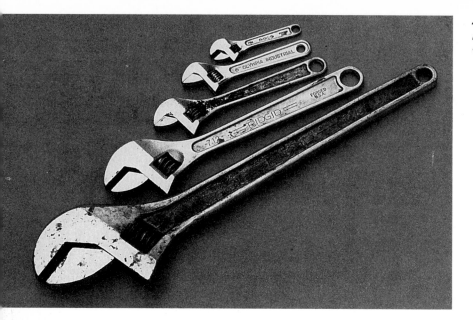

A set of adjustable wrenches is indispensable for performing a wide variety of plumbing tasks.

ADJUSTABLE WRENCHES

Adjustable wrenches have many uses in rough plumbing. For the occasional plumbing job, you could probably get by with just two wrenches, an 8-in. wrench and a 12-in. wrench. If you are planning a large-scale project, it makes sense to have additional sizes. My toolbox contains a 4-in., a 6-in., an 8-in., a 12-in., and an 18-in. adjustable wrench.

The 4-in. wrench is very handy for reaching hard-to-get-at, no-hub coupling bands when a longer wrench will not fit in confined spaces and for working on water-heater pilot-light generator nuts and tube nuts, and other thermocouple connections. It is also the perfect tool for turning on and off my acetylene tank. The 6-in. wrench gripping a 20-in. screwdriver sometimes helps to break loose a frozen screw when I remove pipe straps in remodel work. The 8-in. wrench is used mostly in finish plumbing, for example, to install copper thread by sweat adapters in tub/shower valves.

The 12-in. wrench is useful when working with no-hub pipe, to break off little jagged pieces of pipe on uneven cuts (see p. 90). It is also used in the demolition phase of remodeling work, for breaking down threaded joints in old fixtures. The 18-in. adjustable wrench is sometimes used to remove wood between shallow saw cuts when there is no swing room for the hammer. It is also handy when you have to "crack loose" and snug up water-meter union nuts when removing and replacing the meter in remodel work in order to get a fully drained-down system, as well as for removing the threaded plugs in cleanout fittings.

SOCKET WRENCHES

My socket-wrench set includes the following: a 15-in. long mechanic's ½-in. square-drive ratchet wrench; a #40163 SK mechanic's ½-in. by 20-in. extension; a #40162 SK mechanic's ½-in. by 10-in. extension; a #5418S Proto $\frac{9}{16}$-in. socket to grip nipple backouts; and a #NT 1222 Napa $\frac{11}{16}$-in. socket to grip ½-in. pipe plugs. These tools, used in conjunction with the Ace EX-7 nipple backout (see the discussion on the facing page), allow you on remodel to remove pipe plugs and nipples in tight, deep areas or areas where close-in swing room for wrenches is nonexistent. When the two extensions are joined together and the socket is on the end, the complete assembly is 32 in. long. This arrangement has been an absolute life-saver.

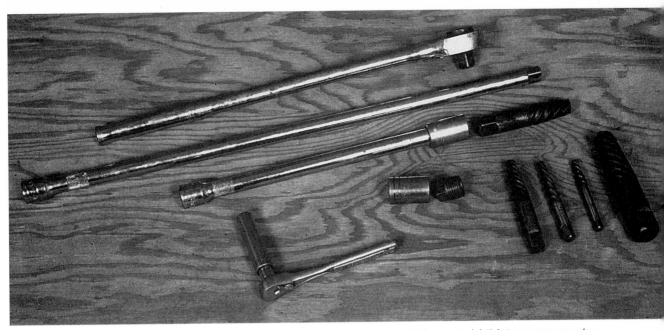

A socket-wrench set with extensions and various sockets and a nipple backout (shown at right) let you remove pipe plugs and nipples in barely accessible areas.

PIPE-NIPPLE BACKOUT AND PIPE TAPS

The pipe-nipple backout (also called an extractor) can remove severed pipe from tees when you are splicing in new water or gas branch lines in situations where conventional wrenches can't be used. There are a number of designs and brands, but I have found the Ace line to be the best. I suggest that you buy the Ace EX-7 for ½-in. pipe. If you should happen to need a larger size, buy only the size that you need.

I also have a collection of pipe taps sized from ¼ in. to 1½ in., which I use for chasing out imperfect threads on new female threaded fittings. The ½-in. and ¾-in. taps get the most use.

MANUAL CUTTING TOOLS

Assembling a plumbing system involves cutting a lot of pipe. There are hand and power tools that will accomplish the task, and I use both, depending on the pipe material and the situation. Manual cutting tools are especially useful where access is limited or the vibration of a power tool might cause damage. Common manual cutting tools include hacksaws, aviation snips and diagonal-cutting pliers.

Pipe taps clean out imperfect threads on new female threaded fittings.

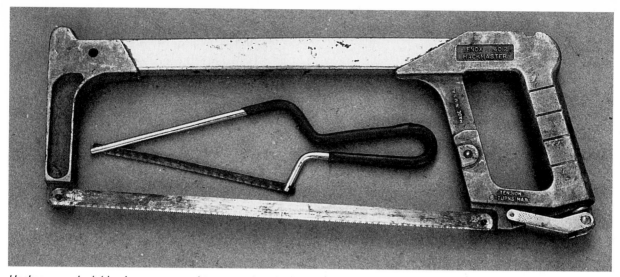

Hacksaws and mini-hacksaws are used to sever pipes in remodel work because they vibrate less than reciprocating saws.

HACKSAWS

Hacksaws come in handy in remodel work. I often use them to sever sections of old galvanized water lines for splicing in tees where the use of a high-vibration tool like a reciprocating saw could cause waterborne sediment, rust and scale to clog faucets, tub and shower valves and toilet ballcocks, not to mention dishwashers and washing machines. Good-quality hacksaws — there are many satisfactory models — have a rugged, rigid frame and a well-designed mechanism for maintaining proper blade tension. If you have several new, good-quality blades of various tooth counts, hacksawing need not be drudgery. A sharp blade, a can of oil and no more than 52 strokes per minute can take you through many, many pipes without ever wearing out a blade.

The inexpensive mini-hacksaw is an indispensable tool. I use it primarily to cut copper and galvanized pipe for splicing in tees in remodel work. Its fine-tooth blade is best employed with teeth positioned for cutting on the back stroke. It vibrates hardly at all, and it gets in tight spots.

AVIATION SNIPS

I use my aviation snips for cutting flat stock for heat venting, for cutting wire lath in stucco when blasting holes for heat-duct thimbles and for cutting galvanized and plastic plumber's tape for hanging pipe.

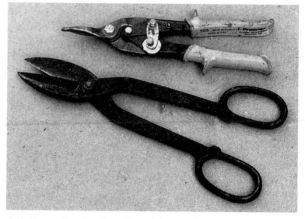

Aviation snips (top) and straight-cutting snips (bottom) come in handy for cutting flat metal stock, wire lath, plumber's tape and strapping.

Most plumbing-supply houses sell three versions of aviation snips with color-coded handles. A yellow-handled tool is for cutting in a straight line, but can also trim wide arcs. A green-handled tool is for cutting tightly arced right-hand cuts; a red-handled tool is for cutting tightly arced left-hand cuts. I carry all three, but find that the yellow-handled tool can do almost any cut I need. I also have a large pair of traditional straight-cutting snips resembling scissors whose longer blades do a good job of removing the old heavy strapping used for hanging iron pipe.

DIAGONAL-CUTTING PLIERS (DIKES)

My 8-in. dikes are made by Diamond Tool Co. I use them mostly in remodeling for pulling the little lath nails left on stud walls that will really open up your flesh if you don't see them. I also use them for cutting metal lath in stucco walls that is too heavy to be snipped by my yellow-handled aviation snips. Another 8-in. dikes that I like very much has an arc on the flat side, which gives your hand good clearance to make flush cuts. This tool is made by Klein Tools.

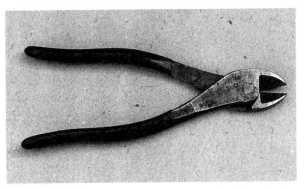

Diagonal-cutting pliers (dikes) can pull out nails and cut through heavy-duty metal lath.

POWER CUTTING AND DRILLING TOOLS

Plumbers really work their power saws and drills. I know of no other trade that is as hard on these tools, so durability and reliability are critical qualities to look for, even though good tools do not come cheap. Spending nearly $300 for a hand-held drill motor and close to $100 for a drill bit sounds outrageous, but buying the toughest and most adaptable tool is cheaper in the long run than having to replace several low-budget models. In this book I mention the brands and models that have served me well in my work; you may have other favorites.

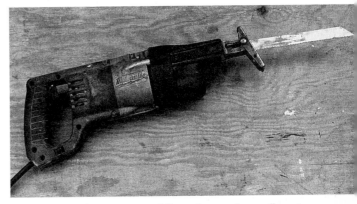

The reciprocating saw can cut through wood as well as various piping materials; shown here is the Sawzall, made by Milwaukee.

RECIPROCATING SAWS

Plumbers use the reciprocating saw more than any other trade. Not only do we cut wood with it, but we also attack all forms of piping — copper, plastic, terracotta, iron, steel and lead. On average, I wear out one reciprocating saw every three years. I have tried every brand there is, and have concluded that no one saw has all the best features. For my latest saw, I came back to the Milwaukee Sawzall.

For plumbers, the #6508 Dual Range, Trigger Speed Control Sawzall is an excellent choice. One of the most difficult pipe materials to saw is iron. It can be extremely brittle and glass-hard, and needs to be sawn very slowly. With the trigger speed control, you can slow down the blade until it strokes no faster than you could comfortably saw by hand. A two-speed saw or one with six to ten preset speeds cannot match the effectiveness of the #6508 when it comes to sawing iron. With two-speed and multiple preset-speed reciprocating saws, I have spent nearly two hours and used as many as eight blades just making one cut through a particularly stubborn 4-in. iron sewer pipe. With the #6508, that chore takes maybe a half-hour and two or three blades.

For the reciprocating saw, there are blades of various designs and levels of quality. Cheap one-piece blades with heat-treated teeth are all right for cutting wood, but not at all good for metal. Bimetal blades, with teeth made of a much tougher metal than the bendable part, are much more useful in the heavy-duty cutting a plumber has to perform. I generally use Lenox blades, but most brands of bimetal blade are acceptable.

JIGSAW

I use a jigsaw to cut a hole in the floor for the toilet flange. This hole should have as little clearance as possible around the hub of the flange because the holes in the top flat rim will split out of the subfloor if they are too close to the edge. The more wood you can leave past these holes, the better off you are, and the jigsaw makes a much better, more controlled cut than the reciprocating saw. Since this job isn't very demanding, any little two-speed, light-duty saw works just fine.

RIGHT-ANGLE DRILLS

The workhorse of the plumber's toolkit is the right-angle drill, which is designed to make holes in the center of a standard stud space of 14½ in. (16 in. on center). My Milwaukee Hole Hawg (#1675) has a two-speed (300 rpm and 1200 rpm) drill motor. The higher speed is for making small holes (up to ¼ in. in diameter) in sheet metal and wood, and the lower speed is for making large holes for piping in both metal and wood. The right-angle drill can be used with various bits (see the sidebar on pp. 11-12).

The right-angle drill bores holes in metal or wood and fits into a standard 14½-in. stud space. Shown here is the Hole Hawg, made by Milwaukee.

Milwaukee is not the only company making a big right-angle drill for rough plumbing, but its Hole Hawg is king of the hill in performance, versatility and popularity. This drill motor is a real bruiser (you cannot use it for any great length of time without getting a bruise). Its gear reduction is such that if you are drilling with a large bit and tie it up, say, in a hidden metal hanger or strap, you are going to go for a ride. But you need all the torque that this drill delivers when you are putting a 4⁵⁄₁₆-in. dia. hole through a double plate on top of a 1⅛-in. plywood subfloor.

CORDLESS DRILLS

I have two cordless drills, and each is more effective in certain applications. I use a Porter-Cable cordless drill (#850) to install the no-hub couplings when I assemble no-hub piping systems. I use a Milwaukee cordless drill (#0398-1) to screw pipe hangers and strapping in place, and drive self-drilling sheet-metal screws into single-wall conduction pipe when installing heater vent systems. An extra battery is a must if you are going to use a cordless drill in earnest on the job.

DEMOLITION HAMMER

The electrically powered demolition hammer is used in remodel plumbing to break up concrete, for example, to get through a concrete foundation wall so that you can bring pipe through. Because this tool costs more than $1,000, you might just want to rent or borrow one when you need it.

SPECIALIZED TOOLS

Cutting and joining sections of pipe are a big part of a plumber's work, and these tasks may require specialized equipment, depending on the materials and techniques being used. These specialized tools are described in the appropriate parts of Chapter 4. Plastic pipe is joined with cements (see pp. 60-61); copper pipe is generally soldered (see pp. 65-68); steel pipe is threaded (see pp. 79-80); and both iron pipe and terra-cotta pipe are joined with no-hub couplings (see pp. 88-89).

Hole saws, auger bits and self-feed bits are used extensively in plumbing (see the photo at right). Hole saws, as the name implies, cut the circumference of a hole; bits bore the entire hole. Auger bits have a lead screw that is an integral part of the bit, and if the screw threads get very dull or damaged, you have to toss out the bit. Self-feed bits have a lead screw that can be replaced or adjusted for extra length.

HOLE SAWS

Hole saws cut in both wood and metal, and plumbers use them primarily in the top and bottom plates to extract nails embedded in wood that's being bored with auger or self-feed bits. Embedded nails can ruin your day by damaging your bits. My hole-saw collection closely matches my Milwaukee drill-bit set for the Hole Hawg. When you encounter a nail while boring with augers or self-feed bits, you simply grab your hole saw and continue working until the nail is sawn away. Then if necessary, resume boring with the drill bit.

Hole saws are sold individually or in sets (indexes). My newest hole-saw index is a Milwaukee, #49-22-4065, but the hole saws are too short to cut all the way through a 2x4. The last hole-saw index that I bought was a Sandvik, which could manage this chore without any problem.

Cutters for the right-angle drill include (left to right) a Milwaukee 2¼-in. dia. hole saw, a Lenox 2¼-in. self-feed bit and a Milwaukee 2⅛-in. self-feed bit.

AUGER AND SELF-FEED BITS

Several companies manufacture drill bits for plumbing, but the bits I own are made by Lenox and Milwaukee. I have a Milwaukee drill index that contains seven bits, sized to cut holes in wood for ⅜-in. up to 2-in. dia. iron pipe. The first four are auger-type bits, which means that I have to push on them to make the hole. The three largest bits are self-feed, so little or no shoving is required.

Installing the fresh-water distribution system usually calls for a lot of boring. For these small-diameter pipes, the Milwaukee bits are all augers. Each Milwaukee auger is long enough to bore through a triple plate or double plate and subflooring, or through triple studs. For boring small holes, I'd destroy my back using only the auger bits. For drilling through single or double plates, I use the shorter Lenox bits, which are self-feed over their complete size range. These bits are especially helpful when boring upper plates while on a ladder.

Lenox and Milwaukee bits each have their pros and cons. The Lenox self-feed bits pull themselves faster through the wood than the Milwaukee self-feeds. However, the Milwaukee self-feeds have cutting teeth around the bit's perimeter, which make for cleaner holes and less splintering when the bit comes out the back side of the material being bored. The Milwaukee bits tend to bore a straighter, if slower, hole through thick material. The Milwaukee self-feeds can bore plates when as much as one-third or more of the bit extends off one edge of the plate. Trying to accomplish this task with a Lenox bit can bind up the drill and split out the plate.

EXTENSIONS

There will be times when the shank of the bit is not long enough to penetrate the material you need to bore, and you will need an extension for the bit. As a rule, one manufacturer's extension will not accept another manufacturer's bit, for there is a difference in the configuration of flats on the shank of the bit. Even with the same manufacturer, this flat configuration can change for different bit sizes. You can usually get by with a 5½-in. bit and an 18-in. extension for each brand.

It's best not to rely upon only one product line to accomplish all the work you need to do. A case in point is drill-bit extensions. Milwaukee uses an archaic, unsatisfactory setscrew technology for holding bits to the extension, but it does make wonderful, slow-cutting, large-diameter bits. Lenox makes an excellent extension that is hexed to mate with its hexed drill-bit shanks, as well as faster (but not as smooth-cutting) self-feed bits with more self-feeding action (in fact, you have to slow them down).

On rare occasions, I have to dig out my 36-in. long ship's augers. One is a ¾-in. dia. bit and the other is a 1-in. dia. bit. I bought them for $1.50 apiece at government surplus over 20 years ago. I don't know what I'll do when they finally break.

SAFETY, COMFORT AND CONVENIENCE

Plumbing a house can be dangerous work. You are lifting and steadying heavy pipes, often in cramped quarters. You are working with noisy power tools in a moist, dimly lit environment. You may also be installing hookups for gas appliances. No discussion of plumbing tools would be complete without mentioning the items that allow you to accomplish these tasks safely and comfortably.

PROTECTIVE GEAR

Eye protection is essential. When you work with a reciprocating saw or a right-angle drill, falling debris is a constant nuisance, and if you're not careful, it can become a threat. Pick a type of eye protection that you are comfortable with — that way, you'll use it. Options include goggles, safety glasses and face masks. I generally use safety glasses with side shields rather than goggles, which tend to fog and scratch easily. (I also wear a dust mask and soft, conical ear plugs when the work calls for it.)

Hand protection is a must for plumbers because of where we put our hands when we work and because of the noxious chemicals in fluxes, lubricating oil and sealants. These days, many plumbers wear disposable gloves, which are sold in drugstores and at health-supply outlets. If your skin reacts badly to the powders in latex gloves, you can buy unpowdered latex gloves. Another choice is disposable vinyl gloves.

The knees are another part of the body that is subjected to wear and tear. I once came close to a leg amputation due to an infection acquired from kneeling on contaminated floors while installing toilets. Since then I have worn knee pads. In addition to preventing contact with dirty floors, knee pads are also kind to old bones. I like the felt-lined, real leather variety. Knee pads are usually sold at lumberyards, hardware stores and builder's emporiums.

Plumbing is hard on the back. Lifting heavy lengths of iron and terra-cotta pipe and threading lots of big pipe can really strain your muscles. Once it was unheard of for "real" men to admit a compromising physical condition, but now, thanks to the feminist movement, vulnerability is nothing to be ashamed of, and the back support (worn on the outside) is almost high fashion.

A small flashlight strapped to the head puts the light where you need it, leaving both hands free.

FLASHLIGHTS

I have tried every major brand and type of flashlight and have yet to discover one I consider well suited to a plumber's needs. However, there are two types of light that you must have: a head lamp and an adjustable spot. It's also nice to have a more powerful light source to illuminate a larger area.

Plumbers often need both hands for working, and it is usually impossible to set down a flashlight and adjust its beam to illuminate the work effectively. A head lamp is the answer. My head lamp is simply a small flashlight attached to a self-fastening strip that wraps around my head. The adjustable spot, which is freestanding, also lets you direct the light beam at your work. It may be needed for long periods of time when working in darkened areas before power is installed in that part of the building, or for jobs where it is impractical to string a power cord. I like the Sportsman and Pasco brands, both of which are all metal and use a 6-volt storage battery with screw-on battery connectors.

An illuminator lets you throw good light over a large area, such as a crawl space, without having to drag electrical cords around. I like to use a fluorescent lantern or the Coleman #5370C1900.

CHAPTER

2

SYSTEMS AND MATERIALS

There are four types of piping systems in residential construction: drain, waste and vent (DWV); fresh-water supply and distribution; fuel-gas supply; and gas-appliance venting. Each system uses different types and sizes of pipes, though you usually have a choice of various code-approved materials for a given application. This chapter focuses on the two essential systems, DWV and fresh-water supply and distribution, including piping materials, and concludes with a discussion of the pros and cons of various straps, hangers and types of pipe insulation used when hanging pipes.

The other two systems are installed only if the structure will have gas appliances, such as a cooktop or range, furnace or water heater. In many areas of the country, the fuel-gas piping system is installed by residential plumbers, but elsewhere this job may be performed by a specialist instead. I design and install fuel-gas piping systems, and I am going to explain how it's done, but that will come later, in Chapter 7. Designing and installing vent systems for gas appliances will be discussed in Chapter 8.

The DWV system is installed first because it is a gravity system (without pumps or pressurization), so it needs the most direct pathways, free of all but es-

sential changes in direction. Also, DWV pipes are the largest-diameter pipes that plumbers install, and the water-distribution and fuel-gas systems, whose pipes are much smaller in diameter, can usually fit around them. Because water-distribution and fuel-gas systems are pressurized systems, they can accommodate more twists and turns without much consequence.

DRAIN, WASTE AND VENT

The DWV system takes away the liquid and solid wastes from toilets, bidets, tubs, showers, washing machines, lavatories and kitchen sinks. Drain, waste and soil pipes run from the individual plumbing fixtures to a septic system or a municipal sewer; vent pipes provide air to the DWV system, and in conjunction with traps, prevent the back siphoning of wastes.

On p. 16 is a schematic drawing of a generic DWV system, with the major components labeled. The fittings that join DWV pipes are discussed in detail on pp. 17-21. Depending on local codes, the pipes could be ABS or PVC plastic, DWV-weight copper or no-hub iron (see pp. 21-24). Some localities allow mixed materials to be used; others do not.

DRAIN, WASTE AND SOIL PIPES

Drain, waste and soil pipes perform similar but not identical functions, and their definitions overlap somewhat. A *drain pipe* carries liquid waste (soapy water) and solid waste (fecal matter, sanitary products and food by-products, including garbage-disposer discharge). A *waste pipe* conveys only liquid waste (waste that is free of fecal matter). By this definition, a lavatory drain pipe is technically a waste pipe, since it normally receives only liquid wastes. However, as a rule, shower and laundry lines are not considered waste pipes because they might convey fecal matter from bathing or diaper laundering. A *soil pipe* handles the discharge from toilets, urinals and bidets. My major code says that a soil pipe may handle just these fixtures or other fixtures connected to it as well.

The main vertical pipes of the system, be they soil, waste or vent, extend through one or more stories and are called stacks. Secondary pipes, called branches, are joined to the stacks with the appropriate (code-approved) fitting. The *vent stack* passes up through the roof of the building. The *soil stack* conveys the waste of the house's fixtures to the *building drain,* which is the biggest drain in the house. The building drain ends at the inside perimeter wall of the building, where it joins the *building sewer.*

Technically, the building sewer is not part of the DWV system. In many areas, the plumber has to have an additional permit to install piping outside the perimeter of the house, including a sewer line. We will discuss this process, however, for those needing to lay their own sewer line (see pp. 96-97 and 127-128).

TRAPS

A trap is a curved section in a fixture's drain pipe that retains water (the trap seal). The purpose of the trap is to prevent the gases that build up in sewer systems from rising back into the house. Methane, one gas found in sewers, is explosive. When plumbing was in its infancy, drainage systems had no traps, and it was not uncommon for structures to blow up and burn. Many of the explosions and fires were blamed on faulty gas lighting and heating, since the dangers of methane were not well understood.

There are two common types of traps: P-traps and S-traps. *P-traps* are used in conjunction with *trap arms* for drains that come out of the wall, which is the usual case these days. S-traps were used on drains

that come out of the floor; they are not allowed by many local codes in new construction today. For all fixtures except toilets, traps must be installed when the fixture is hooked up (see "Sink Wastes and Traps" in *Installing and Repairing Plumbing Fixtures,* the companion volume to this book). The water that sits in the toilet bowl funtions effectively as a built-in trap, so no separate trap is needed.

VENTS

A *vent* is a pipe located along a fixture's drainage system beyond the trap that rises up to the roof and provides air to the system. This column of air helps keep the water in the trap and also allows the liquid and waterborne waste to drain out of the system quickly and efficiently by preventing an airlock. (Think about what happens when you hold your thumb over a drinking straw that's full of water — the water won't drain until you "vent" the straw by removing your thumb.)

There are three commonly used types of vents: individual vents, back-vented vents and loop vents. *Individual vents* rise from the drainage piping through the roof, without joining to any other pipe. *Back-vented vents* are attached to another properly sized vent pipe for convenience. *Loop vents,* sometimes called island vents, are used under island sinks where there is no wall in which to run conventional individual vents (loop vents are not shown in the drawing on p. 16 but are discussed on p. 107).

Wet vents are vents that also serve as drains. In general this is not a good idea (I've done only three such installations in the last 20 years), but in some instances it may be your only choice. Wet vents require the prior approval of your local inspector.

FITTINGS

Fittings join pipe sections together, either to make turns (change-of-direction fittings), to join intersecting pipes (branch fittings), to link segments in straight runs (couplings) or to allow the system to be unclogged (cleanouts). Except for the couplings used to join no-hub iron (see p. 20), fittings are generally of the same material as the pipes being joined. Because fittings are so critical to the functioning of the DWV system, fitting selection for a particular application or location is often specified by code. For specifics on fitting selection and installation, see Chapter 4.

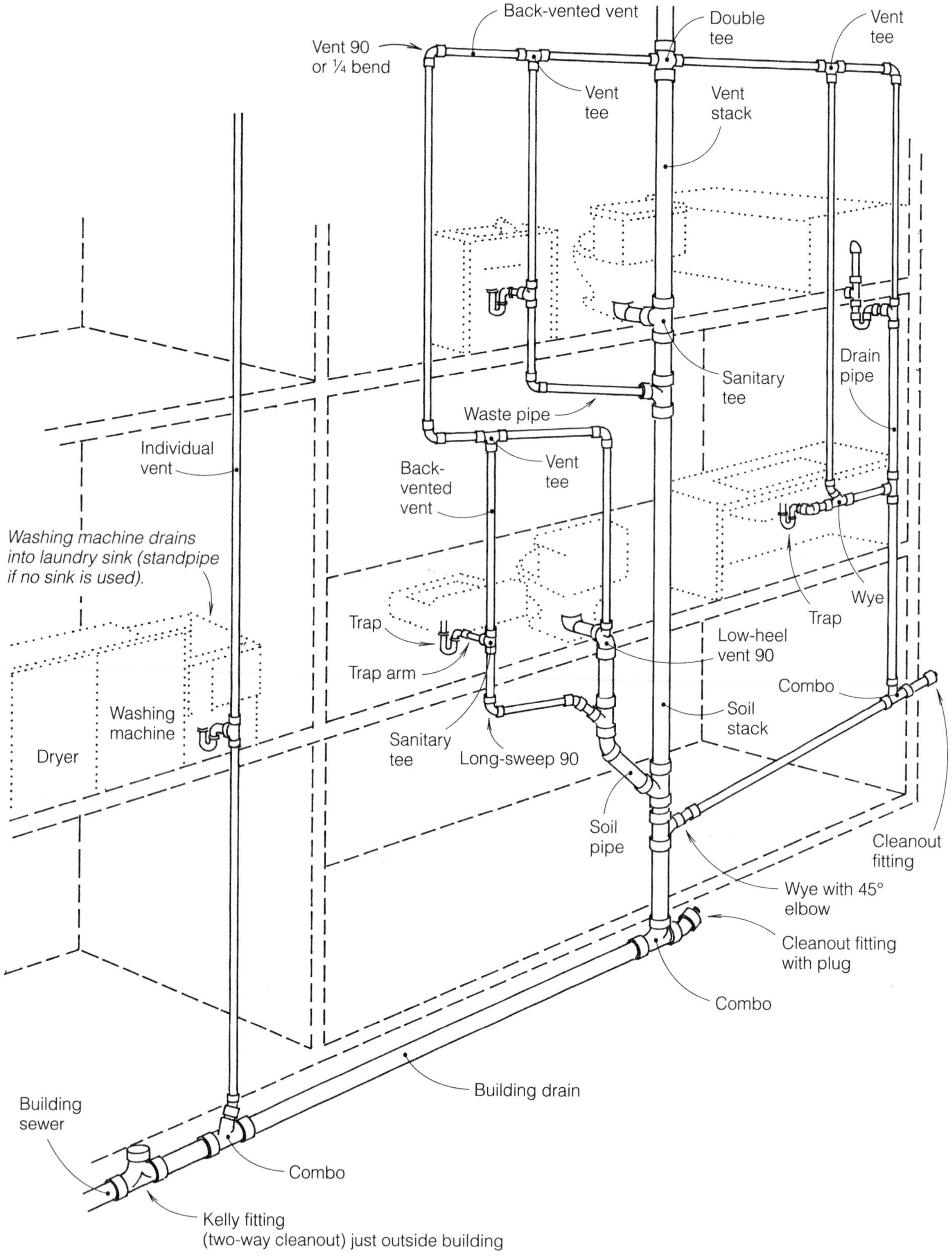

Back-vented vent

Double tee

Vent tee

Vent 90
or ¼ bend

Vent tee

Vent stack

Vent tee

Sanitary tee

Drain pipe

Waste pipe

Back-vented vent

Vent tee

Individual vent

Wye

Trap

Washing machine drains into laundry sink (standpipe if no sink is used).

Low-heel vent 90

Combo

Soil stack

Cleanout fitting

Trap

Trap arm

Sanitary tee

Long-sweep 90

Dryer

Washing machine

Soil pipe

Wye with 45° elbow

Combo

Cleanout fitting with plug

Building drain

Building sewer

Combo

Kelly fitting
(two-way cleanout) just outside building

CHANGE-OF-DIRECTION FITTINGS

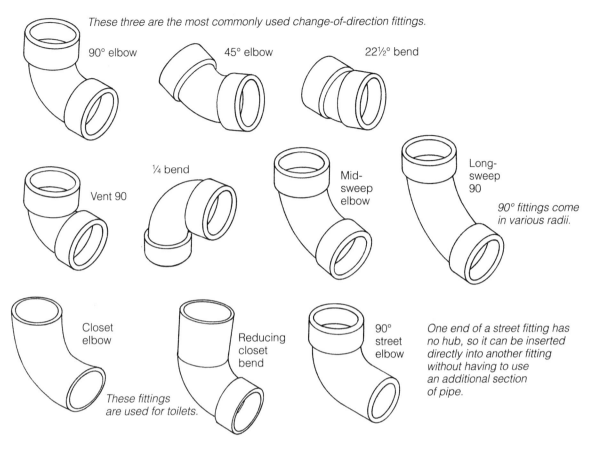

These three are the most commonly used change-of-direction fittings.

90° elbow 45° elbow 22½° bend

Vent 90 ¼ bend Mid-sweep elbow Long-sweep 90

90° fittings come in various radii.

Closet elbow Reducing closet bend 90° street elbow

These fittings are used for toilets.

One end of a street fitting has no hub, so it can be inserted directly into another fitting without having to use an additional section of pipe.

CHANGE-OF-DIRECTION FITTINGS   The only change-of-direction fittings you are likely to need are 90° elbows, 45° elbows and 22½° bends (elbows and bends are interchangeable terms). Sometimes these fittings are described fractionally: ¼ bend (90° elbow), ⅛ bend (45° elbow) and ¹⁄₁₆ bend (22½° bend). Except for no-hub iron, all DWV piping materials use hubbed fittings. In ABS, PVC and copper, change-of-direction fittings can be purchased with one female end (with a hub) and one male, or streeted, end (without a hub). These street fittings save space, because you can insert the streeted end directly into the hubbed end of another fitting instead of having to insert a short section of pipe between fittings. Street fittings are very handy where there isn't much room to work.

There is a choice of radius for 90° elbows, depending on the material. ABS, PVC and copper 90° elbows have the tightest radius, and these **vent 90s,** as they are called, are used only for vents. The next sizes,

which are found in all DWV materials and are all drainage fittings, are the ¼ **bend**, **mid-sweep elbow** and **long-sweep 90.** (The 45° elbows and the 22½° bends are made only in drainage fittings.)

A **closet elbow** is a plastic or no-hub iron 90° fitting used under a toilet instead of a ¼ bend or larger-radius 90° elbow. Its discharge leg is longer than its inlet leg, and it comes in two configurations: standard (discharge and inlet of the same diameter) and reducing (discharge of smaller diameter than inlet). Incidentally, the reducing closet bend is the only code-sanctioned fitting in which the discharge end is smaller than the inlet. The **reducing closet bend** in no-hub iron comes in handy when space is limited, as in shallow (2x8 or 2x10) upper-story floor joists; the plastic version isn't that much smaller than the standard. However, not all companies make identically shaped closet bends.

BRANCH FITTINGS

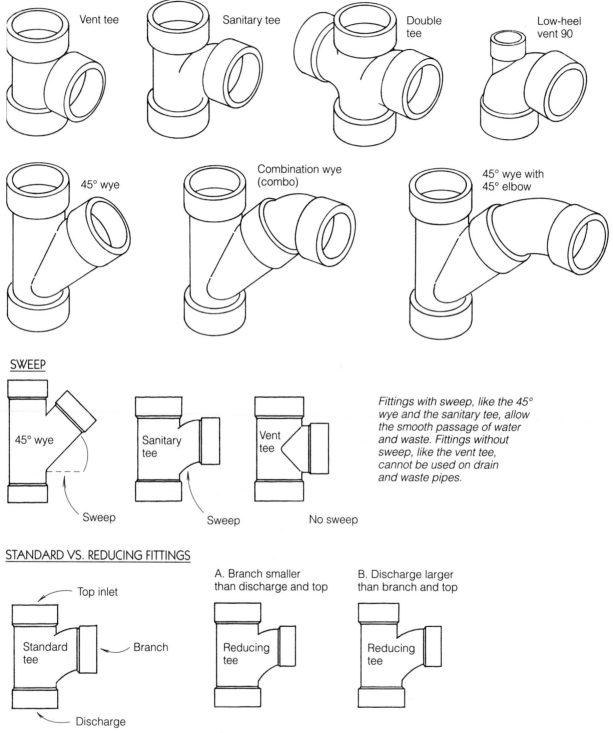

Vent tee

Sanitary tee

Double tee

Low-heel vent 90

45° wye

Combination wye (combo)

45° wye with 45° elbow

SWEEP

45° wye

Sweep

Sanitary tee

Sweep

Vent tee

No sweep

Fittings with sweep, like the 45° wye and the sanitary tee, allow the smooth passage of water and waste. Fittings without sweep, like the vent tee, cannot be used on drain and waste pipes.

STANDARD VS. REDUCING FITTINGS

Top inlet

Standard tee

Branch

Discharge

A. Branch smaller than discharge and top

Reducing tee

B. Discharge larger than branch and top

Reducing tee

On standard tees, all openings are the same size.

On reducing tees, either the branch is smaller or the discharge is larger than the other openings. Reducing-tee sizes are specified in the following order: discharge, top inlet, branch. Thus a 3x3x2 tee has a 3-in. dia. discharge and top inlet and a 2-in. dia. branch.

BRANCH FITTINGS   Branch fittings allow you to connect intersecting vents or drainage pipes. The two basic types of branch fittings are tees and 45° wyes. (There are also 60° wyes, but they aren't used much.) Most codes, regardless of story height or foundation type, want intersecting pipes to be linked with a *45° wye* where space allows.

When selecting a branch fitting for a particular application, you have to consider the issue of sweep. Sweep is the angle or arc in a pipe that allows water and wastes to flow smoothly along. *Vent tees* have a branch without any sweep; *sanitary tees,* which are used mainly to connect trap arms to waste stacks and their vents, have downward sweep on the branch.

In ABS and PVC you have two fittings used only for venting purposes: a vent tee and a vent 90. There are no fittings without sweep in DWV copper and no-hub iron. There are no 4-in. dia. vent fittings in any material, so in many parts of the country, you are allowed to use sanitary tees for venting purposes. A sanitary tee used for venting should be positioned so the branch channels the flow of air downward along the path it will take from the vent opening on the roof as it heads downstream in the drain line.

The *low-heel vent 90,* which is actually a branched change-of-direction fitting, is used for venting and draining a toilet (usually on an upper floor) whose drain runs vertically inside a wall directly in back of and below the fixture. At the center of the fitting's longer outside radius is the heel, and the vent inlet positioned on the heel will always be of a reduced diameter. The top inlet of the 90° elbow accepts the toilet's trap arm, and the bottom discharge is installed inside the wall.

Another useful branch fitting is the *combination wye (combo fitting).* This fitting is a 45° wye with a 45° bend on the end (inlet) to the branch, which allows the branch to connect to the stack at a 90° angle. If you don't have a combo fitting, you can approximate one by joining a *45° elbow to the inlet of a 45° wye.* In many cases the homemade solution is more desirable because you can twist the elbow slightly to compensate for intersecting piping that may not be at a perfect 90° to the fitting's branch. There are also branch fittings with four openings, such as the *double tee,* but they are rarely needed in residential work.

Branch fittings may be either standard or reducing. On standard branch fittings the discharge, top and branch inlets are all the same size. Reducing branch fittings come in two configurations. The discharge and top inlet may be one size and the branch smaller, or the discharge may be large and the top inlet and branch smaller. The sizes of reducing branch fittings are described by calling out the openings in a particular order: discharge, top inlet, branch. For example, a 3x3x2 tee is one with a discharge of 3 in., a top inlet of 3 in. and a branch inlet of 2 in.

COUPLINGS   Couplings join sections of pipe in a straight line; the standard couplings for plastic, copper and no-hub iron are shown in the photo below. Couplings used in drainage usually have both openings the same size. There is no such thing as a downstream reducing coupling because it is illegal to reduce the diameter of a downstream drain line, except for the reducing closet bend (see p. 17). However, you can increase pipe size as you move downstream, and for that you use an increasing adapter.

Plastic (top) and copper (middle) DWV pipe sections are joined with sleeve-like couplings; no-hub iron pipe (bottom) is joined with a no-hub coupling.

No-hub iron pipe and fittings are joined with what is generically called a no-hub coupling (a number of companies, such as Fernco and Indiana Seal, manufacture them). Common variants on the no-hub coupling include the Mission coupling and the CIT coupling. For a detailed discussion of when these couplings are used and how they are installed, see pp. 90-93. Terra-cotta pipe can be joined with Calder couplings, which are discussed on pp. 94-95.

NO-HUB COUPLING

A no-hub coupling is a rubber collar surrounded by a stainless-steel band with a captive worm-drive hose clamp on each end (see the photo on p. 19). No-hub coupling bands are grooved across their width. When the hose clamps on the grooved band are tightened to the design tension, one end of the band nests on top of the other end, with grooves meshed. The mated grooves give the coupling a good alignment and a good portion of its strength.

A Calder coupling can be used to join terra-cotta pipe for buried connections or to join pipes of different materials.

MISSION COUPLING

A commonly used coupling that resembles a no-hub coupling is the Mission coupling, made by the Mission Clay Company. (This is a brand name, not a generic term like no-hub.) The Mission coupling, shown in the photo above left, differs from the no-hub coupling in the design of the outer stainless-steel band and in some cases in the screws used to draw it together. The Mission coupling has a smooth outer band and in some larger sizes uses two straight machine-thread bolts and nuts, pulling through two lugs on its edge clamps to

The heavy-duty, all-rubber CIT coupling may be required for no-hub pipe that is buried in the ground. (Photo by Bill Dane)

draw itself together. Mission couplings are sometimes specified in some localities for buried no-hub pipe connections; they are also used to join DWV pipes of different materials.

CIT COUPLING

There is also an all-rubber coupling that has an all-stainless worm-drive hose clamp on each end (see the photo above right). This heavy-duty coupling, called a CIT coupling (CIT stands for cast iron to cast iron), may be required in some localities for no-hub pipe that is to be buried.

Different types of couplings are used with different piping materials. For ABS and PVC, the coupling is merely a sleeve; the ends of the pipes being joined are inserted and butted inside.

For copper, standard couplings come with interior stops. One type has a full interior ridge, and this type is the strongest. Another type has one or two little dimples stamped through the fixture to form a small lump on the inside wall.

No-hub iron pipe sections can be joined with rubber couplings that clamp together with metal bands. These connectors, which include no-hub couplings, Mission couplings and CIT couplings, are discussed in the sidebar above.

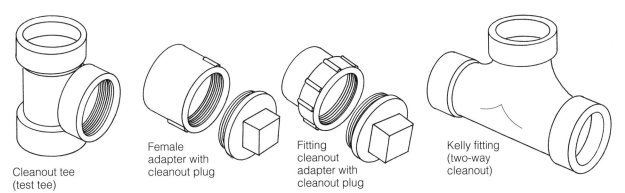

Cleanout tee
(test tee)

Female
adapter with
cleanout plug

Fitting
cleanout
adapter with
cleanout plug

Kelly fitting
(two-way
cleanout)

CLEANOUTS Cleanouts are fittings with removable plugs that allow DWV systems to be unclogged in the event they become plugged up. Codes specify the frequency and locations of cleanout fittings in a system. My major code calls for a cleanout at the upper terminus of each horizontal drainage line and every 100 ft. or fraction thereof on any horizontal piping exceeding 100 ft. in length. An additional cleanout is required for each aggregate change of direction exceeding 135°.

I install cleanouts at the upper terminus of sink drains, and tub and shower drains where practical. Under sinks, *cleanout tees* may be installed above or below the sanitary tee; the threaded plug that seals off the branch can be removed for access to clogs. The branch of the cleanout tee is shorter than the branch of the fixture tee, which in some locales may also be acceptable as a cleanout fitting.

With plastic piping, I use one of two fittings as an upper-terminus cleanout. A *female adapter with cleanout plug* is used on a cleanout extension pipe run. A *fitting cleanout adapter with cleanout plug* can be installed directly in a branch inlet. The fitting adapter has a streeted discharge that slips into the inlet of the branch fitting.

To be useful, cleanouts must be easily accessible. Often, they aren't. That is why I always install a *Kelly fitting* (two-way cleanout) right outside of the house at the juncture of the building sewer and the building drain. An additional advantage of using a

Kelly fitting at the lower end of the building drain is that you aren't required by code to have an upper-terminus cleanout.

Nevertheless, I think it is a good idea to have cleanouts for small, horizontal, 2-in. dia. under-slab drainage lines for lavatory and kitchen sinks, and for laundry sinks at their upper terminals for both slab and off-the-ground foundations. These cleanouts should be vertical tees (test tees) with a threaded branch and plugs, installed above or below the sanitary tees (see p. 106).

DWV PIPE MATERIALS

Several materials are acceptable for DWV piping today: plastic (principally ABS and PVC), copper and no-hub iron. Terra-cotta (extra-strength vitrified clay) was once a popular material for sewer pipes; nowadays it is used less frequently. You might, however, encounter terra-cotta in remodel or repair work, so it is discussed at the end of this section, in the sidebar on p. 25.

Threaded galvanized steel is another material you might encounter in your remodel work. In the past, it was commonly used for vents and interior drains. It is still acceptable for use in these applications, but it is rarely used because of its high cost and time-consuming installation when compared to plastic and copper. However, if you for some reason you want to to use galvanized steel for vents, working with this material is discussed on pp. 78-86.

Because the DWV system is a gravity system, DWV piping can have a lower pressure rating than water-distribution piping made from the same material. For example, copper DWV-grade piping is a lot lighter in weight than Type K or L copper, which is used in pressure applications. That's why it is strictly forbidden to use DWV-grade materials for any fresh-water distribution and supply.

PLASTIC (ABS AND PVC)

Plastic piping has largely replaced metal in many parts of the country because of its low material and installation cost. Nowadays you also find it used for fresh-water distribution (see pp. 31-33) and fuel-gas yard piping (see p. 165), as well as DWV.

The most common plastic DWV pipe materials in use today are acrylonitrile butadiene styrene (ABS) and polyvinyl chloride (PVC). Different geographical areas usually specify the use of one or the other; seldom can you take your choice. You can generally find these materials anywhere; even in communities where they are not sanctioned by the local authority.

PVC was developed in the mid-1930s. The basic raw material used in its manufacture is acetylene or ethylene. From these materials, a family of plastics (polyvinyl chlorides) may be formulated to be either soft and rubbery (shower curtains), hard or rigid, or of any degree of hardness in between. Rigid or un-plasticized PVC is the form of the material that is generally used for pipe and fittings. Initially it was expensive and difficult to produce. As technology improved, it became cheaper and more readily available.

ABS piping first came into use in the 1950s in oil fields and in the chemical industry. In 1959, a DWV system constructed of ABS plastic was installed in a test house (Research House Number One, built by prominent Arizona builder John F. Long) in Phoenix. Two years later, a buried section of the building drain was excavated, cut out and forwarded to a test lab. It got a clean bill of health. In 1984, another piece of the drain was tested and again given good marks. The use of ABS drainage piping was approved by the FHA in 1960, and since then, it is estimated that more than 2 billion feet of ABS pipe have been installed for this purpose.

ABS and PVC pipes have several advantages over metal pipes: ABS can be cut easily and quickly with a special handsaw, and PVC up to 2 in. in diameter can be cut with the same saw or with a tool called a plastic scissors (see p. 59). These plastics are soft materials that debur well; they are also lightweight, and they are readily available. Once pipe and fitting are cemented, the joints are very resistant to leaks due to vibration and movement, so they are especially well suited for kitchen sink-waste systems with a garbage disposer. The pipes also have a very smooth bore, which is an advantage with solid waste.

Joinery is a snap with ABS or PVC plastic pipe and fittings (see pp. 54-62). In contrast to copper, which must be soldered, there is no danger of starting a fire when joining ABS or PVC pipe and fittings. In contrast to no-hub iron fittings, which require a larger space to work in, ABS and PVC fittings can often be joined to pipe when the fitting is actually buried in multi-layered plates and in other extremely confined areas. This ability to assemble in less than perfect framing conditions is the biggest reason for using plastic pipe and fittings for DWV.

Plastic DWV pipe also has its negative aspects. ABS, should it catch fire, burns with a vengeance and produces extremely toxic fumes. PVC does not burn quite so savagely, but it also produces very toxic fumes. This is why most communities that sanction plastic pipe restrict its use to residential structures of not more than two stories. Solid and liquid wastes flowing through plastic drain and waste pipes are very noisy. And when used horizontally, plastic pipe is prone to sagging unless it is supported on very close spacing (every 4 ft.). ABS comes in two versions, solid wall and foam core; nowadays, the latter is more common. Solid-wall pipe is solid material all the way through, but foam core has a thick inner layer of ABS foam with a thin layer of solid material on each side. As you might expect, foam-core ABS is more susceptible to damage, especially from solvents in the cement used to join it.

Plastic pipes are more easily damaged than iron or copper. Drywallers often drive their sharp little nails into it (this is not a big deal in vents because the nails seal so tightly that no gases will leak out). You also have to watch out for other hard, pointed objects like screws, pick points and shovels. Of all DWV pipes, plastic is the least resistant to damage from overzealous drain-cleaning contractors. Some industrial-strength, liquid drain-cleaning agents can badly deform and even melt ABS pipe and fittings. And when plastic pipe is left unprotected in direct sunlight, it is weakened; exposed piping should always be painted or wrapped.

DWV plastic pipe should not be bent, which builds stresses in the pipe wall and causes failures. Expansion and contraction of the structure can also cause breaks and leaks in plastic piping if you do not allow for this movement when you install the pipes.

Both plastics are messy to work with, but PVC is worse because you have to use a runny primer as well as the cement. There are published cautions concerning the safe application of this primer for use with PVC. The cement used to join ABS has an equivalent agent but is part of the formula for the cement. Many people suffer headaches from the glue vapors of these products, and pregnant women, especially, should not breathe the fumes at all. When I have to work with plastic pipe, I wear latex examination gloves on both hands and have plenty of rags nearby to wipe off drips and runs from the pipe and fittings. A hat keeps drips out of my hair.

Plastic pipes come in two weights: Schedule 40 and Schedule 80. Lighter-duty Schedule 40 plastic pipe comes in 20-ft. lengths. ABS pipes are usually a dark charcoal grey to black in color and there will be a printed description on the side, giving the name of the manufacturer, schedule and designated application: DWV. Schedule 40 ABS plastic fittings carry similar information on their sides, in raised letters. Schedule 40 PVC pipe and fittings for DWV are usually white or cream color. They will also have printed information on the pipe and identifying marks on the fittings.

Schedule 80 PVC pipe used for buried main building supply will come with a belled end; this heavy-duty pipe starts at 2 in. I.D. Schedule 80 PVC pipe is also available threaded at both ends in sizes from ½ in. through 2 in. I.D. Unthreaded Schedule 80 PVC pipe can be threaded with both hand and powered equipment using dies made especially for plastic. There is no Schedule 80 ABS pipe.

ABS (top) and PVC (above) are the most common DWV pipe materials used today; these plastic pipes are relatively inexpensive and easily joined with simple tools and minimal skills.

COPPER

There is a copper pipe specified for drains, waste and vent (Type DWV) and a companion weight of fittings. DWV copper pipe differs in several ways from copper pipe for fresh-water supply. The DWV copper-pipe wall is considerably thinner. DWV copper fittings are lighter in weight than those used for fresh-water supply, and their sockets are shallower. And DWV copper is considerably less expensive, though hardly cheap. Occasionally in my repair and remodel work I find houses that have 3-in. and 4-in. copper drains and vents, which would never be specified today. Small-diameter DWV copper is comparable in materials cost to no-hub iron but in assembly time (in most situations) it is the slowest of all possible choices, and therefore the most expensive. That is why few homes are DWV plumbed with it today.

Nevertheless, copper pipe has its advantages in the DWV system. If you can afford to use it for a kitchen-sink drain line, you will have fewer stoppages than with other materials. I recommend running a copper kitchen-sink drain line as far as possible toward the main building drain/sewer and insulating the drain line. That way the hot water heats up the copper and grease and soaps remain liquid all the way to the 4-in. building drain, where they are flushed out quickly. The same advice applies to the shower drain, but again, only if you can afford it.

Another use for DWV copper is on tall vertical and exposed vent runs, where trying to hold up no-hub iron long enough to get it secured to the building would pull your arms out of their sockets. DWV copper also looks a lot better than no-hub iron. Copper pipe of a given I.D. size is smaller in O.D. than no-hub iron, and the wall thickness of the fittings is less, so you get a more streamlined appearance. Also, the smooth outside of the pipe and fittings look much better when painted.

Because of its smaller outside diameter, I use DWV copper where structural integrity is a consideration (for example, if I have to bore a lot of holes down and up a stud wall or external bearing wall) or in a tight space (for example, turning a horizontal vent run 90° inside a channel or getting an upturn in a shower vent in the space of an 8-in. floor-joist height). DWV copper is used in 1¼-in., 1½-in. and 2-in. vent lines and an occasional 1½-in. or 2-in. drain line. It's a

DWV copper pipe and fittings are rarely used today, except in certain limited applications because of the time it takes to solder them together.

good choice where its light weight, small fittings and outside pipe diameter allows it to get it into places where plastic or no-hub won't fit. In these special applications, the better performance of the material and the preservation of the structure's integrity are worthwhile trade-offs against the cost.

Copper pipe is joined by soldering (see pp. 69-71). One of the more time-consuming aspects of installing DWV copper is that you have to figure out the orientation of the pipe and fittings in their finished position, set them down at some other location and solder them together (to avoid starting a fire or scorching some portion of the structure, especially the exterior). Then you reinstall them, leaving an open fitting to which the next piece can be soldered.

NO-HUB IRON

No-hub iron pipe used to be the traditional choice for DWV piping, and even though it has been largely supplanted by plastic today, it still has several advantages. No-hub systems can be easily modified, since the couplings can be undone and then quickly retightened. You'll be glad you chose no-hub iron if your inspector recommends changes in the system — not an infrequent occurrence.

However, there are some disadvantages. No-hub iron is a very heavy material (it comes only in 10-ft. lengths). Also, the fittings for no-hub 2-in. pipe can pose a problem because they are considerably longer

Terra-cotta pipe and fittings are sometimes spliced in to repair old sewer lines. The pipes and fittings come from the factory with a Calder coupling on one end.

Terra-cotta (extra-strength vitrified clay) is one of the oldest pipe materials known to man, and it's still around — in some parts of the world there are still functioning terra-cotta drains that were installed by the ancient Romans. I don't know of any engineers or architects who specify terra-cotta nowadays, but there might be some builders who still use it on new construction. It is quite an economical pipe.

If you get involved with remodel or repair plumbing, you will probably have to deal with terra-cotta pipe and fittings. You will either be digging up and tearing out old leaking pipe runs and replacing them with no-hub iron or plastic, or repairing the leaky pipes by splicing in new fittings.

In the good old days, terra-cotta pipe for residential use had a male end (spigot) and a female end (bell). It was joined by simply packing mortar around the spigot end after it was inserted into the bell. Nowadays the pipe comes from the factory with a Calder coupling on the barrel and on any branches. Calder couplings are simply larger versions of the all-rubber CIT couplings used to join no-hub iron (see the sidebar on p. 20). They are discussed more fully on p. 93.

Of all the pipe materials used in residential construction, terra-cotta has the thickest wall. To adapt terra-cotta to other piping materials, you merely add a rubber bushing (designed for each particular material) to the rubber coupling before introducing the new pipe material. Bushings for terra-cotta to iron, terra-cotta to plastic and terra-cotta to copper are easy to get, but a bushing for terra-cotta to asbestos cement pipe can be as hard to find as a black pearl. I always check first with my suppliers before cutting any pipe.

As terra-cotta ages, it changes color and character and presents different challenges. New terra-cotta is a rich yellow-orange or red-orange color. After years of burial and exposure to water and earth it often turns to a dark chestnut brown. Sometimes it turns olive or yellow-green. After many, many years in the ground it can take on a dark lavender hue. Good-quality terra-cotta will not deteriorate with age under normal conditions, but it can't handle shocks and flexing. It is never used inside the house or in exposed applications outside.

Today, the most common length of 4-in. dia. terra-cotta pipe is 5 ft. (because of its weight, it would be difficult to carry in longer lengths). Years ago, in the San Francisco Bay area, there was a fair amount of 3-in. pipe and fittings used, but today I have a difficult time finding any 3-in. pipe and fittings to make small repairs, so I have to use 4-in. iron or plastic instead.

New terra-cotta pipe can be cut fairly easily with the soil pipe cutter. New terra-cotta pipe "pops" quickly, leaving a very clean edge that is also razor sharp (watch your fingers). Old terra-cotta pipe is a lot more difficult to cut cleanly. After years of doing battle in the trenches, you develop a sixth sense that helps you make this decision: "pop" it or saw it. For more on cutting terra-cotta pipe, see pp. 94-95.

No-hub iron was the material of choice for DWV piping before the advent of ABS and PVC piping.

than those for ABS (see the photo on p. 98). This extra length means that you have to route your drains and vents for tubs, showers, laundry standpipes and lavatories through more-central stud bays so you have adequate space for the sweeping bends. For example, you may have to elongate the holes through standard 2x4 upper and lower plates on the bottom of the plates with the reciprocating saw before mid- and long-sweep no-hub 90s will lie plumb in them. This sort of work can be very time-consuming (and therefore expensive).

Some types of no-hub drainage pipe come from the factory with their inside walls coated with asphalt. Before installation, check the pipe for pieces of asphalt coating protruding from the inside wall. These mounds can hang up solids on their journey out of the building and cause repeated stoppages. If you find any asphalt mounds, you should install the pipe so that they are on the top arc, where they will cause less trouble. Obviously, asphalt mounds don't matter in venting.

FRESH-WATER SUPPLY AND DISTRIBUTION

Fresh-water distribution supply and piping carries water to all areas of the structure where fixtures require it. This system operates by pressure. The pressure may be supplied by a local water utility's pumps when the water is transported to your neighborhood through street mains or by the house's well pump.

In some areas, a distinction is made between supply and distribution piping and service piping. (Supply and distribution refers to the structure's main hot and cold supply pipes and the branch lines off them going to fixtures and appliances; service piping carries potable water from the water meter or well to the building.) In such areas, it is not unheard of for plumbing contracts to specify that the rough plumber install only the supply and distribution piping. However, plumbers commonly find themselves installing the service pipe as well, so the installation of both systems is discussed in this book.

In residential systems, the biggest safety concern has been keeping contaminants out of the drinking water. The fresh-water system should have no cross-connections to landscape watering systems or waste systems. As fresh-water sources dwindle, communities in the not-so-distant future may have treated-water utilities to supply homes with recycled water for toilet flushing, landscape needs and possibly laundering. If we reach this point, there will be even a greater need for diligent attention in the installation of separate supply systems for recycled and potable water.

The supply and distribution system is fairly simple (see the drawing on the facing page). **Risers** and **branch lines** transport the water from the outside source, fittings join sections of pipe, and valves turn the water on and off. As shown in the drawing, the hot- and cold-water lines terminate just outside the wall, as a plumber would leave them at the end of the rough-plumbing phase. Later, the plumber returns to the structure and installs the sinks, toilets, bathtubs and other fixtures, making the necessary connections to the water-supply and drainage pipes. That phase of the job, finish plumbing, is the subject of this book's companion volume, *Installing and Repairing Plumbing Fixtures*.

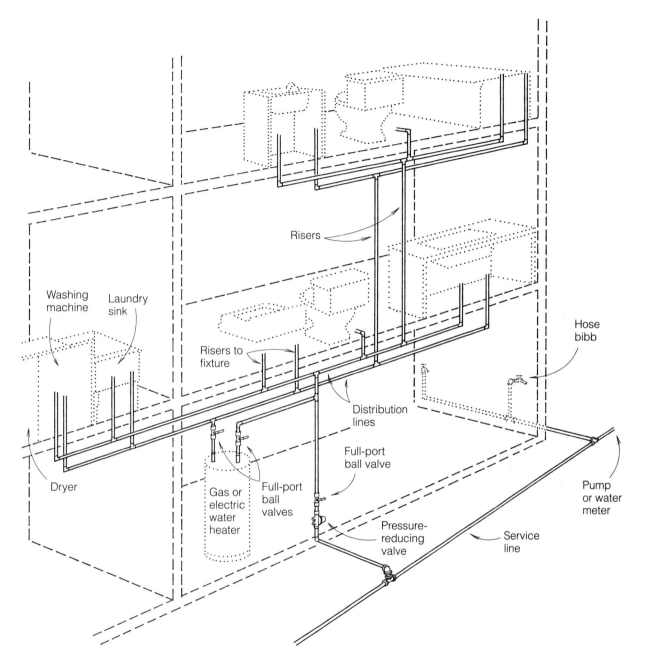

Risers

Washing
machine

Laundry
sink

Hose
bibb

Risers to
fixture

Distribution
lines

Full-port
ball valve

Dryer

Gas or
electric
water
heater

Full-port
ball
valves

Pump
or water
meter

Pressure-
reducing
valve

Service
line

RISERS AND BRANCH LINES

In water-distribution jargon, a **riser** is a pipe that carries water from one story to the next story above. The term is also used for the short vertical pipes that bring water from the branch lines to the fixtures. **Branch lines**, both hot and cold, branch off the main in-house supply and run to individual fixtures.

FITTINGS

Water-supply fittings connect pipes. In fresh-water piping, there are no vents or cleanouts, so the connections are fairly simple. Change-of-direction fittings include tees, 90s and 45s, and other connectors. Specific fittings for various applications are discussed in Chapter 4, under the material being joined.

VALVES AND REGULATORS

Valves regulate the flow of water through the system and allow you to shut it off completely. Three main types are used in residential plumbing: gate valves, ball valves and globe valves. **Gate valves** have a regulating mechanism that resembles the lift gate in ancient irrigation systems (lifting the gate releases the water dammed up on the upstream side). In a gate valve, the gate is made of brass and is lifted up by a male threaded stem that meshes with female threads that pass through the gate. Turning the valve handle counterclockwise threads the gate upward, allowing water to pass underneath. Turning the handle in the opposite direction forces the gate downward, restricting the flow or shutting it off completely.

Ball valves have a ball in the middle instead of a gate. The ball has a hole through the center that allows the water to flow through; on **full-port ball valves**, the hole nearly matches the inside diameter of the pipe. All ball valves have a lever handle that swings through 90°, turning the water on and off much more quickly than a gate valve and also allowing you to see whether the valve is open or closed from a considerable distance.

I use ball valves in most of my plumbing work, now that their price is comparable to that of gate valves, since full-port ball valves have almost no frictional loss (see p. 44). Both gate and ball valves are offered with ports to be soldered to copper pipe or threaded onto steel or rigid plastic pipe. Some models have "barbs" for insertion and use with flexible plastic pipe such as polybutylene and polyethylene.

WATER-SUPPLY VALVES

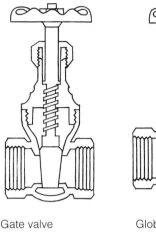

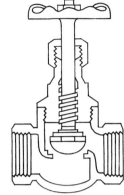

Gate valve

Turning the handle lifts the gate, allowing the water to pass through.

Globe valve

Water changes direction inside valve.

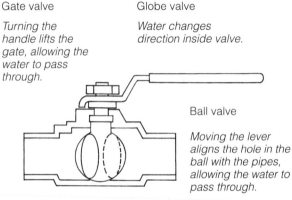

Ball valve

Moving the lever aligns the hole in the ball with the pipes, allowing the water to pass through.

Globe valves have a washer at the end of the stem that fits into the valve seat to stop the flow of water. Because the water changes direction inside the valve, globe valves are notoriously inefficient, and I never use them.

Valves that turn the water on and off include the angle stops installed under sinks and toilets, in-the-wall tub and shower valves, hose bibbs and washing-machine valves. Additionally, there are valves installed on water heaters that regulate temperature and relieve pressure. On systems with water pressure over 80 psi, a pressure-reducing valve is installed on the water distribution main. The regulator is offered with both solder connections for copper pipe and female threaded connections for threaded steel and plastic pipe.

SUPPLY AND DISTRIBUTION PIPE MATERIALS

For fresh-water piping, the code-approved choices are copper, galvanized steel and various kinds of plastic. Some regional codes permit the use of both metal and plastic pipes in one system.

When selecting a piping material, you should consider longevity and durability as well as cost. I often do repair work in homes built in the early 1900s and before, and their original galvanized-steel water piping has held together well. How long will it last? Who knows? How long will copper-fresh-water supply piping last? Again, who knows? It hasn't been around long enough or as long as steel piping. Carpenters drive a lot of nails through my copper piping, but this rarely happens with galvanized steel.

The mineral content of the water supply is also a factor in pipe selection. You want to use a pipe material that will best transmit the water for the longest period of time with the smallest amount of mineral deposits, which restrict water flow. Scale buildup is more of a problem in metal pipes than in plastic pipes. If your water tends to build deposits inside pipes, polybutylene or polyethylene pipes might be a better choice than copper or galvanized steel. Your local inspector will be able to advise you on the best choice for your area.

COPPER

Inside the structure, and specifically inside the walls and under the floor, I like to use copper water-supply pipe. It has a fast assembly time, and it can be placed in much tighter confines than can galvanized steel. But because it has relatively thin walls, care must be taken to protect it from nails (see p. 141).

There are three types of rigid copper tube (pipe) used in residential plumbing: Type K, Type L and Type M. Type K, which has the thickest wall, is the heaviest, while Type M is the thinnest and lightest. The color of the printed description on rigid copper tube indicates the wall thickness of the pipe. Type K is green. Type L is blue. And Type M is red.

Type M is the standard choice of most communities for use inside the building, off the ground. Some communities allow Type M to be fastened to the outside of the building and even buried in the ground.

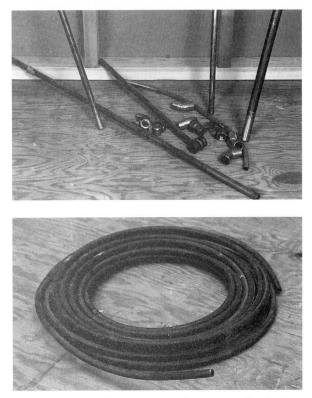

Copper is an excellent choice for fresh-water distribution piping. It comes in two forms: rigid pipe (top) and rolled tubing (above).

Most communities designate Type L for burial, when allowed. For the most part, you don't see much use of Type K in residential work. I use a little of it at the termination points of water lines where lack of space would make it difficult to support thinner pipe adequately. Rigid copper tube is sold in 20-ft. lengths.

Types K and L copper also come in rolled "soft" versions, in 60-ft. long coils. Soft copper has not been annealed, which accounts for its flexibility, and costs twice (or more) what rigid copper costs. Something that costly is used sparingly when there is an alternative. But in some applications soft copper is worth the extra cost. One such application is laying the building's service line from the meter to the structure. Soft copper can be unrolled in the trench very quickly, and it conforms to the unevenness of the trench bottom. A lot of labor is saved by not having to sweat fittings to rigid tube to make it lie in the trench at the proper depth.

Another benefit of using soft copper in the ground is having a longer continuous length of copper without couplings in it. The fewer the joints in the line, the stronger it is — a very important consideration in climates where winter and summer temperatures vary drastically. Mother Earth pulls and shoves with remarkable force between the extremes of freezing and sweltering. Rigid copper tube in 20-ft. lengths requires two soldered joints (the weak links) to achieve the same 60-ft. distance as one length of soft copper. When a water line has to be run under a concrete slab, it's far better not to have any joints. Soft copper can pay for itself in faster labor times and as insurance against very expensive demolition and repairs.

In the San Francisco Bay area, where I work, copper water-supply pipe began receiving wide acceptance in the late 1950s and early 1960s. At that time, most communities that allowed the use of copper for burial insisted on Type K. After about ten or so years though, most relaxed their standards and allowed Type L to be buried instead because the material fared better than they had expected.

One word of caution. As more copper pipe and fittings are imported into this country, it is harder to find suppliers willing to sell the U.S. product lines. The imports, especially fittings, tend to be lighter in weight with shorter radiuses and have more defects. I always try to avoid the imported no-name fittings and rely instead on known manufacturers like Streamline, Mueller, Nibco, Lee and U.S. Brass.

A potential problem with copper is that it requires fire to be joined, and as everyone knows, fire and wood can be a dangerous combination. I always carry some pieces of sheet metal to protect combustible surfaces from my flame, and if I'm going to be sweating a lot of copper I keep a fire extinguisher handy as well. (For more on soldering, see pp. 69-71.)

Electricians and telephone-wire installers ground their wiring to the plumbing pipes. With galvanized-steel pipe this is not a big problem. But copper is so electrically sensitive that if the grounding clamps are plated iron and steel, after a while they can start corroding the pipe. If any ferrous metal (electrical boxes or various straps and supports) touches the copper and the pipe has a grounding strap on it, the pipe won't last as long as it should. For more on metal corrosion, see *Fine Homebuilding* #62, pp. 64-67.

At high water pressures, copper pipes tend to be noisier than galvanized-steel pipes, but the noise can be moderated or overcome by some extra care in reaming the pipe's bore on assembly and by the use of certain insulated hangers and bushings. Working with "pressure" copper pipe and fittings is basically like working with DWV copper. For more on working with copper, see pp. 64-77.

GALVANIZED STEEL

Galvanized-steel pipe has been used for fresh-water supply for about 90 years, both in the ground and inside the building. It has now largely been supplanted by copper pipe, but it is a good choice in some applications. The only place I use galvanized steel for water is in the ground; in this application my major code requires the use of protectively coated (plastic-coated) pipe. I use it most often as a main building water supply between the meter and the structure if the line happens to be on or very near the surface of the ground, where it will be subjected to vehicular traffic. Here, the tougher steel will outlast the more vulnerable copper and plastic pipes.

My biggest complaint about the material is the poor quality of pipe and fittings now being sold in our country. Most of the galvanized-steel and "black" (unplated) steel pipe and fittings used in the building industry comes from Asian countries, and the quality of the imported material is inferior to the products that used to be made in this country.

Galvanized-steel pipes should not be used in areas where the water is high in mineral content. In such areas, new pipe can clog with minerals and rust out in a few years. If you are considering using galvanized-steel pipes, it's a good idea to ask a number of local plumbers whether galvanized steel is a good choice for your area.

Galvanized-steel pipe comes in 21-ft. lengths. (Other rigid plumbing pipe, such as copper, PVC and ABS, comes in 20-ft. lengths. I've never been able to figure out why.) To join lengths of galvanized-steel pipe into a water system, you have to thread the ends, and this takes at least three times longer than soldering copper pipe. (For more on threading pipe, see pp. 82-83.) Another drawback to galvanized wa-

Galvanized-steel pipe and fittings, with ends threaded. Good-quality galvanized piping is sometimes used for fresh-water supply piping buried in the ground.

ter piping inside the building is the difficulty in assembling pipe and fittings in very tight places because of the need for space to swing pipe wrenches and fittings. I marvel at the handiwork of old-timers when I'm under houses making repairs or splicing in new lines. I see where they had to run their piping and I can imagine the hours spent in that little corner when I can have my copper pipe runs installed in a matter of minutes.

PLASTIC

I like to work with plastic water-supply pipe outside the building. Because it is so easy to join, you can lay a lot of pipe quickly. The material is inexpensive, compared to metal piping. It weighs very little and is easy to handle. When a building site sits 200 ft. from the water meter and code dictates a large-diameter line for that run, copper would be far too costly. Even galvanized steel would be very costly, and leak-free threaded pipe of such a large diameter is very hard to achieve in a pressure system. For all these reasons, plastic pipe for water supply seems a logical option; choices include polyvinyl chloride (PVC), chlorinated polyvinyl chloride (CPVC), polyethylene (PE) and polybutylene (PB).

Plastics are relative newcomers to the supply and distribution scene, and they are not sanctioned for the same applications throughout the country. Why one particular type of plastic is or is not allowed is largely a matter of politics. Before you begin planning your water system, you should check with the local authority and find out what is allowed in your area. For example, where I work ABS is strictly a DWV material; it is not allowed for water service. In some jurisdictions, PVC is used for fresh-water service piping to a structure. It is not sanctioned for water distribution within a building, although a related material, CPVC, is allowed by some jurisdictions for water distribution. PE is allowed for water service, but PB is not allowed for service or distribution. I have used PVC and PE for service piping.

Extreme heat and cold can be problems with certain types of plastic pipe. PE and PB have a good resistance to freeze cracking. But you should take extra care not to let water freeze or transmit water above 200°F in PVC or CPVC pipe.

Is plastic pipe a health risk? There have been warnings about possible health risks attributed to one plastic, PVC. However, in the end, it's the local codes that have the final say. If PVC is allowed in your area, you might ask your local public-health department about any new information that could possibly influence your choice of piping. At the present time there is no evidence that PVC contaminates drinking water. One extremely detailed and hard-to-find report on the subject (you might be able to locate it in a large public library) is "Plastic Pipe Installation: Potential Health Hazards for Workers" (Field Investigation F1-88-002 from the California Department of Health Services). According to this report, the greatest danger of working with plastics is direct skin contact with cements and primers and inhalation of their vapors. Protective clothing (including gloves) and ventilating fans would go a long way toward reducing health risks to plumbers who work with plastic pipe.

POLYVINYL CHLORIDE (PVC)   PVC pipe for use in fresh-water supply comes in two weights, Schedule 40 and Schedule 80. Since this material is so inexpensive, I always buy the much heavier Schedule 80 for water-supply lines buried in the ground because of its superior resistance to impact and crushing, even though my code allows Schedule 40. (I use Schedule 40 for DWV because after this system is installed, it is usually protected from impact by walls and floors.) Be very careful when purchasing PVC pipe in the "budget" type home stores, which often sell a lightweight pipe that might not be sanctioned for water service in your area. Anything lighter than Schedule 40 should be avoided at all costs.

Fittings that are glued or cemented to the ends of PVC pipe or to each other are called solvent-weld fittings. There are two liquid agents used in assembling PVC pipe and fittings. One is a primer and the other is the cement. The PVC pipe and fitting material is so tough that the cement by itself will not make a sufficiently permanent, watertight joint. The primer (tetrahydrofuran, or THF) softens the outer surface of the pipe and female surface of the fitting allowing the cement to get a better grip on the two mating parts. For more on joining PVC, see pp. 57-58.

CHLORINATED POLYVINYL CHLORIDE (CPVC)   Some areas allow lightweight CPVC plastic pipe and fittings for the hot- and cold-water distribution system inside the house. I do not install plastic lines in any structure (I use copper instead). The only application I have for CPVC is recirculation lines for outdoor hot tubs and spas where the pump and heater are installed in a separate location from the tub.

Assembly of pipes and fittings is the same as with standard PVC, except that for CPVC you might have a different primer and a different cement (see p. 63). Over the last few decades, CPVC has become a popular choice in areas of the country where the water has a high mineral content because the minerals don't adhere to the walls of CPVC pipe. But because CPCV is more fragile than galvanized steel or copper, it must be installed carefully so it is protected from impact and freezing.

In recent years there have been a number of published cautions concerned with suspected leaching (into house water supplies) of chemicals found in CPVC pipe. If this issue is of concern to you, you might contact your local public-health authority for the latest update.

POLYETHYLENE (PE)   Polyethylene first appeared on the American pipe market in the late 1940s in non-pressure mine-drainage applications. Today, PE pipe is also used for transmission of potable water (pipe approved by the National Sanitation Foundation for this application carries the NSF stamp). This material is now my choice of plastic piping for water service when cost is a huge issue. PE is used outside of the foundation for buried fresh-water service lines up to 2 in. in diameter. PE has all but retired galvanized steel for water-well to house service runs because with PE the distance can be covered in one continuous run without couplings. Because hot water softens this pipe, it is not allowed for water-distribution lines inside the house.

For pressurized fresh water, codes call for either high-density PE (Type 3) or medium-density PE (Type 2). Your local supplier will know which one you should use. Low-density (Type 1) PE is used for low-pressure applications. In residential plumbing, nylon or copper-alloy insert fittings (also called barbed fittings) are generally used with PE pipe. They are shoved into the pipe and secured with a stainless-steel hose clamp. For more on joining PE pipes, see p. 64.

PE pipe is black in color, sized from ¾ in. to 2 in. and generally sold in coils 500 ft. long. It is easily handled, has good flow characteristics and is easily coupled. PE also has good flexibility, high strength-to-weight ratio and chemical inertness. It is light in weight, coilable (easy to lay in a trench) and resistant to corrosion, mechanical shock, freeze/thaw conditions and abrupt changes in pressure.

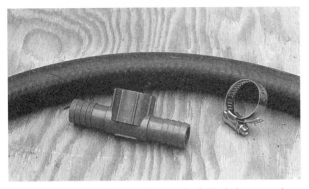

Polyethylene pipe is a durable and relatively inexpensive choice for buried fresh-water service lines up to 2 in. in diameter.

Thermal expansion is not much of a problem with PE. At 73°F, coiled pipe will expand 1 in. per 100 ft. of pipe per 10°F change in temperature. Pipe allowed to straighten will expand almost 6 in. per 100 ft. Because PE pipe is so flexible, snaking it in trenches is an adequate way to compensate for thermal expansion and contraction. By comparison, at 80°F steel expands .61 in. per 100 ft. and copper expands .87 in. per 100 ft.

POLYBUTYLENE (PB)   Polybutylene is a latecomer to the residential-plumbing scene. I first encountered PB pipe in southern California in the early 1970s. For quite a few years the Uniform Plumbing Code allowed PB pipe for fresh-water distribution, both outside and within the building. In those days, for use inside the building, the material had to be fusion welded using expensive, specialized equipment. Outside in the ground, barbed insert fittings and hose clamps, flared joints and mechanical compression joints were allowed. However, PB pipe has recently been dropped from the UPC's list of acceptable piping materials (althouth the UPC continues to publish the installation standard). There seems to have been a problem with the use of one type of ring used to crimp the pipe to fittings. The ring would fail, causing separation at the joint. My local water supplier, which serves almost 1.5 million customers, has stopped using the material.

Not all areas ban PB pipe. Many proponents claim that PB has been unfairly criticized and continue to install it where codes permit. If PB is allowed in your jurisdiction and you wish to use it, I won't try to dissuade you. I would only urge you to take the same time and care in boring pathways for PB that you would for copper or steel, and follow all of the manufacturer's recommendations for its installation. If you do, you should end up with a good system.

PB pipe has some nice qualities. Flow rates for PB and Type K copper are very close, with up to only 1 gal. per minute difference for 1¼-in. pipe. It is similar in flexibility to medium-density polyethylene and has strength properties that fall between polyethylene and polypropylene (polypropylene is a very tough material used in industrial piping applications). It can also be kinked repeatedly without breaking. And there are a lot of adapters for joining it to other types of piping (see pp. 63-64).

Unlike other plastics used for pipe, polybutylene is relatively soft when first extruded. After 24 hours it resembles polyethylene in hardness and strength; after four days it attains approximately 90% of its ultimate strength; in ten days it reaches its final, stable form. It is resistant to freezing (down to -50°F), and also resistant to creep (cold flow), which may be defined as the dimensional change over time of a material under load, over the temperature range of -10°F to 190°F. What all this means is that fittings that have been mechanically tightened should remain tight.

PB pipe is available in 20-ft. lengths and in coils up to 1,000 ft. The coiled pipe has a tendency to resist lying flat when you uncoil it, which makes it difficult to install in straight runs for in-house fresh water. It is easier to use straight lengths of pipe, especially when threading pipe through bored holes in stud wall. The coiled pipe used for burial, for water-service lines and yard piping is a different pipe from the type used for in-house distribution. Your local supplier will be able to help you make the proper material selection.

STRAPS AND HANGERS

Pipe support is a big issue with plumbing inspectors. Well-supported piping systems last longer and are quieter in use. With copper fresh-water distribution piping and a good- to high-pressure water service, it is extremely important to do a good job of fastening the system to the structure. If you don't secure the copper pipe about every 4 ft. to 5 ft., you may end up with noisy, pounding pipes when valves are opened and closed quickly. It can be difficult to figure out which pipes are causing the problem once the structure is completely finished, and if the offending pipes happen to be inside walls, correcting the problem may be frustrating and expensive. (For more on dealing with the problem of noisy pipes, see pp. 142-144). Galvanized fresh-water piping is also subject to shock pounding, but not quite so much as copper. The hangers used to support these piping systems are inexpensive but do take a little more time and effort to secure conscientiously.

For horizontal copper piping, my major code requires support every 6 ft. for pipe up to 1½ in. in diameter. For vertical copper piping of any diameter, my major code requires support at every story. Horizontal galvanized pipe, in my major code, needs support at 10-ft. intervals. That's quite a bit less support than for copper. Polybutylene requires support every 32 in. for horizontal cold-water lines. Because the pipe sags much more when transporting hot water, you should support horizontal hot-water lines every 24 in.

DWV no-hub iron pipe, even 2-in. pipe, needs lots of support because it's so heavy. For supporting vertical pipe, I use halved, bolt-together riser clamps that clamp around the pipe at each floor's bottom-plate level (see the photo above right). These clamps support the weight for that floor-height of pipe, and no clamp then holds more than it should have to. However, when water lines occupy space close to the vertical drain lines, these clamps cannot be employed, and I have to resort to traditional plumber's tape looped around a fitting's branches.

For supporting horizontal iron pipe, there are few options. I use the traditional galvanized-steel or copper plumber's tape. Tape might be less pleasing aes-

A riser clamp supports vertical no-hub pipe at floor level.

thetically than some other hangers but it is a lot stronger and more practical. With some care, you can even make plumber's tape look respectably tidy.

For supporting PB pipe, you have to consider its flexibility and its high rate of thermal expansion. PB expands and shrinks markedly with heat and cold, more than metal pipes, so it's important to give it the space and support it needs. When boring holes in wood through which to send the pipe, make them at least ⅜ in. oversized, and just because the pipe is flexible, do not be sloppy in aligning the holes.

If you cover the pipe with a ½-in. thick protective layer of spongy material (plumber's foam pipe wrap) you can use metal straps to secure the pipe to structural members. On bare PB pipe, you must use only plastic strapping. If you want to do a quicker and visually more pleasing job, use the plastic, one-hole strap called Q Talons. For pipe in stud walls on 16-in. centers, the support factor is no issue, but it is important for pipe on joist bottoms or held flat to vertical surfaces. On vertical runs cold-water lines should be supported at least every 4 ft., and hot-water lines probably closer to every 3 ft. With long, straight runs of PB pipe in stud wall, it is a good practice to run the pipe into a gentle loop in one stud bay. Don't make straight runs and then anchor the pipe at both ends; this can cause leaks from too much tension at the joints of pipe and fittings. PB needs a minimum of 6 in. of slack for every 50 ft. of pipe.

Several companies manufacture specialized plastic hangers for copper and plastic. These usually consist of a pipe-encircling hanging loop and some form of rod or strap that either stabs into a boss on the top or snaps onto an integral flange on the top of the loop; you can usually get an extension section of rod or strap. I don't use these systems because I have seen far too many broken ones when I'm in and under houses doing repairs.

TRADITIONAL PLUMBER'S TAPE

Traditional plumber's tape is a flat galvanized-steel or copper strip, sold in rolls, with a continuous line of punched holes down the middle. The holes are usually one of two patterns: either all the same size or an alternating pattern of large and small holes. The holes on the tape with same-sized holes are usually just large enough to pass a ¼-in. by 20 machine bolt through. Usually the larger of the two holes on the tape with alternating-sized holes will also allow a ¼-in. by 20 bolt to pass through. For hanging 3-in. and 4-in. iron pipes from floor joists, I use the heaviest galvanized-steel tape I can find.

Plumber's tape traditionally was sold in three different thicknesses, but today two are more common: light and heavy. Local authorities used to demand (and in some cases still do) that only the heavier tape be used for hanging iron pipes from floor joists; smaller-diameter pipe could be hung with lighter-weight tapes. Sometimes an extra-heavy-duty punched strap is specified for hanging large-diameter (3-in. and 4-in.) iron pipe. This 8-ft. to 10-ft. long strap is almost twice the width of the rolled tape and probably a full ¹⁄₁₆ in. thick. It may also have either the same-sized holes or alternating-sized holes punched down the center.

For copper pipe in residential construction, there is basically one general weight of copper plumber's tape to use. Genuine 100% copper plumber's tape is very expensive, but well worth the price because there will be no corrosion when it touches the copper pipe. A lot of copper-plated steel plumber's tape is being sold as traditional solid-copper tape. This material rusts quickly when left in moist conditions and is subject to dielectric corrosion when the plating oxidizes away. The hole pattern for the copper tape in

Galvanized-steel plumber's tape (light color) is used for hanging iron pipes from joists. Plastic plumber's tape (dark color) is used for hanging plastic DWV pipes.

my locale is almost all same-sized holes. When you handle either galvanized or copper tape, wear gloves to prevent cuts.

PLASTIC PLUMBER'S TAPE

For hanging plastic DWV pipe, I use plastic plumber's tape, which is very different from traditional plumber's tape. Some of the differences are pleasant, some are annoying. I also use plastic tape for copper pipe that runs parallel to the joists because it won't scratch the pipe and it won't cause the pipe to corrode. I use the plastic tape to support the copper pipe the same way that I use traditional galvanized tape on iron pipe (see pp. 116-118).

Plastic tape is sold in 100-ft. rolls boxed in cardboard; partial rolls of shorter lengths are sold in plastic bags. Buy the 100-ft. rolls — the material is so inexpensive that it doesn't pay not to. The cardboard box, with the aid of a little tape, acts as a dispenser and keeps the coil manageable. Once you open the plastic bag, however, the tape turns into a big, kinking mess.

There are some advantages to using plastic plumber's tape, even of someone of the "old school" like me. All the holes in plastic tape are very small, so you can start a screw through a hole and then stop, leaving the screw held firm. That can be a big help in tough-to-reach areas; you don't have to hold the screw while you try to drive it in with the screw gun. Sometimes you can slide the tape up a vertical surface and, with a tip extension in the gun, bury the screw into the wood. Point the natural curve of the tape into the vertical surface to do this; if it points away, you may not be able to "shoot" the tape (extend it out so you can use it) or get the screwdriver or nut-driver tip on the screw.

On the negative side, plastic tape retains its natural coil. You cannot straighten it out and expect it to stay straight when you let it go; it curves right back. That limits your ability to shoot the tape into tight spots. Because the holes are all very small, you can't use nuts and bolts; you have to cradle the tape every time. If you overtighten a screw or set a nail too hard, the tape has a tendency to split. Very cold weather makes the tape stiff and liable to splitting. Another problem is that plastic tape will catch fire or melt if you get the soldering-torch flame too near.

The holes in plastic tape are not as uniformly spaced as on traditional tape, which sometimes makes it difficult to place a screw in just the right spot. If you try to drive a screw through a solid section, the tape often splits. That's not such a big deal (you just cut another piece), but on an off day in a dark hole it can be tedious. And speaking of dark holes, most plastic tapes are dark, and seeing the holes in a dark working space can be a little frustrating. Perhaps it's a little thing, but there is no folding the material down on itself to hide the screw heads. Black drywall screws are not as obvious, but you must take more care not to overdrive them. The plastic tape is prone to splitting if you use bugle or countersunk screw heads. I like to use the self-drilling, hex-headed bit-tip screws because the flat-sided head evenly pressures the tape to the wood without the threat of splitting. But their cadmium or zinc plating really contrasts with the black tape.

HANGERS FOR COPPER AND STEEL PIPING

A glance at any plumbing catalog will reveal an array of ingenious solutions for hanging and supporting pipes, and new ones continually come on the market. The photo on the facing page shows several hangers that are used for copper and galvanized-steel piping. Like exotic tools, I suspect that some of the more specialized hangers might not still be around in ten years. Time will tell.

SPLIT, TWO-HOLE, FLAT STRAPS   The split, two-hole, flat strap looks like a plastic version of the traditional tinned or galvanized-steel two-hole strap used on galvanized and unplated iron pipe. There is a real difference though. The part of the strap that conforms to the pipe is not just a semicircle with legs bent out from the sides, as on the metal strap; it is a complete circle, with a split wall at the bottom. Through the miracle of plastic injection molding, the split wall is also an integral part of the base. The entire pipe is held off the surface the strap is nailed or screwed to. This allows space for thermal insulation, and it allows the pipe to slip back and forth through the strap as it expands and contracts with changes in temperature. The split bottom lets you pry open the clamp and slip it over and onto the pipe. This is handy when you are by yourself hanging long lengths to floor-joist bottoms. Many companies make versions of this strap and market it under their own trade names, but if you ask for split plastic two-hole straps, you should be able to get what you need.

This strap, which is used on copper or PB pipe, has several other good points. Because the copper pipe is totally isolated from whatever the straps are secured to, they can be screwed onto steel and concrete, materials that would otherwise cause the copper to corrode. Because of the complete circle design, you can slide several hangers down onto a long piece of pipe, and they won't drop off as you crawl under a house with it. The split base enables you to add additional straps after the pipe is partially secured with only a minimum number of them.

Split, two-hole, flat straps are commonly screwed to the bottom edges of floor joists on first-floor construction over a crawl space or basement and on the top of ceiling joists for pipe runs in an attic or false

Hangers for copper and steel piping include the following: a split, two-hole, flat strap (1), a snail (2), a split, two-hole, high-eared strap (3), plastic bushings (4), flat, two-hole, metal straps (5), a J-hook (6), coated wire (7) and a felted strap (8).

ceiling. With these straps I use Malco hex-head bit-tip screws rather than nails because they are quicker to fasten, and the screws let me install the strap much closer to objects than I would be able to do if I had to use a hammer. With hammering you might cause vibration damage to plaster walls, and you might strike the copper pipe instead of the nail head. The bit-tip screw enlarges the nail hole in the base as it passes through without cracking the plastic, and I can move the location of the strap if it is first put up in a temporary location for reasons of soldering logistics. Because the plastic eventually deteriorates in sunlight, I do not use these straps in exposed locations.

SNAILS   Snails, so called because they look like a garden snail in profile, are fairly new; whether they develop a following among plumbers remains to be seen. Snails come with a nail already inserted in position for setting. But because only one nail is used per strap, these hangers cannot be used for masonry or metal surfaces. Snails are probably best suited to floor-joist bottoms and ceiling-joist tops.

SPLIT, TWO-HOLE, HIGH-EARED STRAPS   The split, two-hole, high-eared plastic strap is especially versatile. Like the split, two-hole, flat strap, it can be used for hanging pipe in horizontal runs from floor joists and blocking and for securing pipe to the top of ceil-

ing joists. But the high-eared strap is secured to the face of the joist, not the bottom or top edge. These hangers are especially handy when you are running pipe in tight crawl spaces; you don't have to lie on your back to set them. Also, if you are running lines on top of ceiling joists already roofed in and are also close to the exterior wall or the attic space is really low, you don't need an awkward, vertically held screw gun to set these hangers.

Another nice thing about this hanger is its ability to hold pipe vertically to the back side of a block in a plumbing wall (where you wouldn't be able to get at the screws of a flat strap). The high-eared strap can be peeled open, shoved onto the pipe, slid up or down until it rests on the top or bottom edge of the block, and then screwed in place. I find this strap especially helpful when working vertically, securing tub and shower valve piping to blocks.

PLASTIC BUSHINGS   Plastic bushings are sometimes used to hang pipe. However, these plastic sleeves are more than twice the diameter of the pipe (the difference is the width of the webbing in the bushing) and may also weaken the wooden structure when used at the wrong location. The magnitude of the task of boring lots of large holes will probably be appreciated only by plumbers and electricians, who often have to perform this chore. The issue is time. The larger the hole, the longer it takes and the more sweat it takes. If you have a lot of studs to bore for plumbing walls, using bushings could mean hours of extra labor. The bushings do help reduce noise but using them can also mean having to install more change-of-direction fittings to keep your piping away from other wiring and pipes inside the walls.

FLAT, TWO-HOLE, METAL STRAPS   Flat, two-hole, copper and copper-plated steel straps are a variation on the two-hole, galvanized-steel strap of old. This design holds the pipe very tight to a surface because the pipe does not fit all the way underneath it. A small arc of the pipe sticks out past the mounting feet. When this strap is screwed in place, a lot of compression force can be exerted on the pipe. You have to be cautious when using this strap on hot-water lines. If you nail or screw it tightly to wood members, the pipe may produce a very audible "popping" noise or a noise you can mistake for dripping water when the pipe heats up and expands. I put small pieces of felt under hot-water pipes when I use this strap, and I don't quite tighten the screw all the way down. That way, when the pipe expands, friction won't build up between the pipe and the wood.

Metal straps have some advantages over plastic straps. They are stronger, and in some situations you need strength more than you need pliability. Unlike plastic straps, copper or copper-plated steel straps will not break down in sunlight. I prefer the strap made from solid copper. In some coastal climates, almost anything plated quickly degrades to the base metal, and it's a lot of work to replace corroded hangers.

J-HOOKS   The J-hook is a sharpened steel hook with copper plating (pure copper is too soft to be pounded like a spike, with a hammer). The J-section of this hanger is dipped in vinyl plastic to act as a cushion and insulator. I have seen J-hooks used for both copper and galvanized-steel lines. But put a handful of these J-hooks in a coffee can and leave them outside for a while — they become a lump of rust. I don't use them because it takes a hammer to set them, and hammering causes a lot of vibration.

WIRE HANGERS   Wire hangers for fresh-water copper have an isolating plastic coating, those for steel pipe do not. These hangers offer a speedy means of hanging pipe from floor joists, and when you are coming back to add insulation afterwards (see p. 139), they are particularly convenient. However, these hangers have some drawbacks. They tend to pop loose after they've been installed — the prongs of the hanger just back out of the wood. I have seen many of them hanging by only one leg several years after they were installed. Invariably, they do not pound straight into the joist; one leg folds over to the side and you have to pull them out and try again. When I use them, I do not set the prongs too deep and I use only hangers enough to keep the system suspended. Once the system passes the water test, I insulate the pipes; and on this final phase of the installation, I make sure to install far more hangers than are necessary, figuring that some will fall out. I carry a small pliers to straighten bent prongs so each one is set as best as possible.

FELTED STRAPS   Felted straps are a lighter version of flat, two-hole, galvanized-metal straps. The felt liner dielectrically isolates the copper from the galvanized steel and also sound insulates the pipe from the wood structure. I have always liked the simplicity and strength of felted steel straps, and I like using hangers made from natural materials.

INSULATION

As energy costs continue to climb, pipe insulation on hot-water distribution lines becomes more and more important. Many localities require pipe insulation; your local inspector might even suggest a preferred brand name and catalog number. Even if there are no requirements in your community at the present, it makes sense to insulate the hot-water distribution piping that is not inside insulated space: piping on the exterior of the building, below grade or under concrete slab and piping run in unheated crawl spaces, basements and cellars. Hanging insulated pipes is discussed on p. 139.

There are two very popular types of pipe insulation marketed nationwide for residential construction: fiberglass wrap and extruded closed-cell foam.

FIBERGLASS WRAP

Fiberglass wrap has been available for years. It comes in widths from 1½ in. to 4 in., usually backed by foil, and is packaged in small rolls. Because it is in ribbon form, you must wrap it spirally around the pipe to cover linear distance. When pipe runs are close to wood members or other pipes, fiberglass is very time-consuming and difficult to wrap. In terms of time and cost, fiberglass wrap is a relatively expensive option. It is hard to know how much thermal protection you get with it because the thickness of the wrap can vary greatly, depending upon how much you overlap the edges. Also, you don't get a lot of distance for the amount of insulation in the package. This insulation does have a useful application, though, when used in conjunction with extruded, closed-cell foam insulation (as explained below).

EXTRUDED CLOSED-CELL FOAM INSULATION

Extruded closed-cell foam insulation, made popular by the solar industry, is now economical to use on small-diameter (½-in. through 1-in.) piping. Some hardware stores might stock it in 8-ft. to 10-ft. lengths, but more common in my area are bags containing four or five pieces about 3 ft. long. Plumbing suppliers usually handle the long lengths. The wall thickness will vary and so will the R-rating.

This material is easily installed. For an existing pipe run, the lengthwise slit is peeled open and the material is then shoved onto the pipe. After installation, it will snap back into its original cylindrical shape. You might find a brand that has the slit not cut all the way through, but if you peel it with your fingers, it parts open. This type is especially useful when you want to lay a line below grade or protect a pipe passing through concrete. Instead of peeling it open, you slide it down the pipe as you lay it, and it is less likely to "float off" through an already parted slit or to let moisture in through the slit.

I recently saw another brand of extruded closed-cell foam insulation with a new twist. The full-length Imcoa K insulation, manufactured by Insulation Ma-

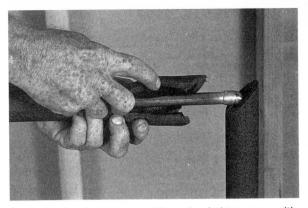

Imcoa K extruded closed-cell foam insulation comes with preglued mitered ends, a convenience when insulating pipe that changes direction.

terials Corp. of America, has a self-sealing, fully cut slit with tear-off tapes. The ends of the insulation are preglued and mitered at 45°. This miter feature is a big help with tees and 90° change-of-direction fittings (see the photo above). Although all closed-cell foam insulation cuts very easily with a pocketknife, making accurate miters on this round material can be a bit exasperating. If you have to run an insulated line through a high-traffic area where everyone will see it and you want a nice-looking job, this product with premitered ends will save you a lot of trouble.

If I'm insulating out-of-view piping in a basement, crawl space or attic with extruded foam and there are a lot of offsets in the run, I don't waste a lot of time and material trying to produce good, thermally efficient miters. I simply install closed-cell foam insulation on the long straight runs and wrap fiberglass ribbon insulation around the short offset sections, using duct tape to hold the ends of the fiberglass ribbon to the foam and keep the joints from unraveling.

Unlike fiberglass, foam insulation does burn, and when it does it releases some harmful by-products. This fact might have some bearing on whether you use it in or close to living space.

CHAPTER
3

SIZING THE PIPE

Plumbers are usually responsible for the design of the plumbing system, and a critical part of plumbing design is sizing the pipe. Sizing pipe is largely a matter of understanding a few simple concepts and meeting or exceeding the minimum conditions stipulated by code. Once sizing is done, you can turn your attention to the layout and installation of the DWV system (Chapter 5) and the water supply and distribution system (Chapter 6). Pipe sizing for gas supply is discussed in Chapter 7.

GENERAL DESIGN CONSIDERATIONS

As shown schematically in the drawing on the facing page, factors that affect pipe sizes in a house include the water volume and pressure, the total water demand (expressed in fixture units), the height of the fixtures above the water supply, frictional loss due to distance and piping material, and the developed length of the pipes (the length along the centerlines of the pipe fittings). Before you design a plumbing plan, you need to take these factors into account. The fresh-water service, supply and distribution system is sized first, based on the building plans for the house and the factors discussed below. Then the DWV system is sized accordingly.

PRESSURE, VOLUME AND FLOW

To work effectively, plumbing fixtures need to be supplied with water at a sufficient pressure. Begin designing the system by considering the pressure in your source of supply. Is it relatively constant or does it fluctuate during a 24-hour period? What is the minimum pressure range? The normal pressure for a street water main is usually between 45 and 80 pounds per square inch (psi). In my area, the city main pressure may vary by as much as 20 psi, depending upon the time of day. Some neighborhoods have peak pressures of 140 psi. Private well systems tend to operate at a lower pressure than city mains; about 25 psi to 40 psi is typical.

Because fresh-water supply and distribution is a pressure system, various regulatory devices may be required to keep things running smoothly. These include booster pumps, pressure-reducing valves, vacuum breakers and expansion-tank and check-valve systems (see the sidebar on pp. 42-43). If you have a

CENTRAL ISSUES IN DESIGNING A WATER-SUPPLY SYSTEM

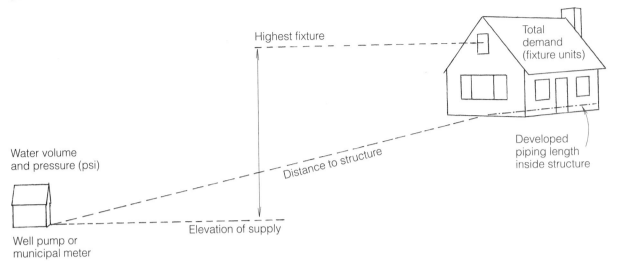

Highest fixture

Total demand (fixture units)

Water volume and pressure (psi)

Distance to structure

Developed piping length inside structure

Elevation of supply

Well pump or municipal meter

well for a supply, you probably won't need a regulatory device. However, if your supply is a city main, then you have to consider the restriction both the meter and any needed regulator will apply to the service and supply pipe (see the discussion on p. 45).

Fixtures also need a certain volume of water to make them function properly. This demand is sometimes called the flow rate. In fresh-water supply piping, water flow is described in gallons per minute (gpm). In DWV piping, flow refers to the rate at which water and waste move down through the system and out of the house. In a fast, easy-flowing drainage system, fittings are arranged so that they direct the flow of effluent down and out of the system with the least interference, as discussed on pp. 14-15.

FIXTURE UNITS

Various major codes assign water-consumption values to each fixture, as an aid in sizing the supply pipes that bring them fresh water and the traps and trap arms that carry away used water and wastes. My major code uses the term "fixture unit" for this value. The more water a fixture uses, the higher its fixture-unit value. For instance, a lavatory is one fixture unit; a kitchen sink is two fixture units; a toilet is three fixture units. Charts of fixture-unit values as they relate to water-supply pipes and trap and trap-arm diameters appear on pgs. 47 and 50, respectively.

The sum total of fixture units represents your building's total demand; that is, what would be required to operate all of the fixtures simultaneously — a worst-case scenario that would probably never happen. So you can often fudge a little bit on the pipe size in some situations.

Fixtures like tubs, showers and sinks need both hot and cold water piped to them. When plumbers are considering only cold water or only hot water for a fixture that has both, they often hedge their bets slightly and go with a lower fixture unit-value for that fixture because the users of the fixture generally mix the water rather than using all hot or all cold. Most plumbers use a 75% figure.

Toilets need only cold water. Other cold-water only items include hose bibs (sill cocks and wall hydrants), automatic-sprinkler/drip-irrigation valves, and water-injection valves for high-tech furnaces. These too have to be tallied into the total demand figure for the building.

If you are plumbing a house supplied by a city main, your total demand figure may be so high that you'll need a larger water meter (and a larger pipe from the city main to the meter). Getting a larger meter is not cheap. At the present time in my area, a ⅝-in. meter for new construction costs about $3,200, and a ¾-in. meter costs about $6,800.

Not every water source will give reliable water pressure straight out of the water main, but various regulatory devices can compensate for deviations from the ideal. Pressure that is either too low or too high can be corrected by installing a booster pump or a pressure-reducing valve; backflow problems can be corrected by installing a vacuum breaker or an expansion-tank and check-valve system.

A pressure-reducing valve is required if the water-supply pressure is greater than 80 psi.

BOOSTER PUMPS

Sometimes an inspector won't allow a house with a lot of plumbing fixtures to be built with the existing low main pressure. In such cases, a booster pump may be called for. A booster pump can raise the pressure of the in-house main and allow the house to be built as planned, possibly with smaller in-house pipe sizes. For unassisted systems calling for larger-diameter (1¼-in.) pipe and fittings, a booster pump can be cost-effective. Of course, you'll need an electrician to provide electrical service for the booster pump. Amtrol manufactures three sizes of residential pressure boosters that are unitized; you don't have to buy separate components and design the system yourself. A midsized booster pump costs about $1,200.

PRESSURE REGULATORS

If the supply pressure to your building is too high (more than 80 psi), you will need to install a pressure regulator, technically called a pressure-reducing valve. There are lots of pressure-reducing valves on the market, but quality varies widely. Well-made pressure-reducing valves are assembled with all non-ferrous or a combination of stainless-steel and non-ferrous body parts, including assembly screws, and contain an integral strainer.

Pressure-reducing valves have a rubber diaphragm and spring at the base of their bell-shaped part (see the photo above); these components reduce pressure in the system by about 25 psi to 90 psi.

VACUUM BREAKERS

In some parts of the country, major and local codes may require the installation of a backflow-prevention device on the water-service line. These devices prevent any siphonage of water back into the building if the main service line is momentarily interrupted or shut down. The big concern with landscape systems is the possible introduction of pesticides and herbicides into the house fresh-water supply by siphonage. A vacuum breaker (also called an atmosphere vacuum breaker) is one such device. Until relatively recently, these devices were used only when irrigation systems were drawing their water from the house supply.

The vacuum breaker (see the photo at left on the facing page) has a chamber that is open to the atmosphere. This chamber is sealed off by a moving part (checking member) that is held in place by water pressure when the service line is under pressure. If the pressure falls away, this moving part drops down and the chamber is opened, allowing at-

Devices that prevent the back siphoning of contaminants into the water-supply system include the vacuum breaker (left) and the check valve (right).

mospheric pressure into the service line. This breaks the siphon action that would draw toxic irrigation water back into the building supply.

Most major and local codes place very specific limitations on the location and installation heights above grade for all backflow devices, including vacuum breakers. Check with your inspector to see if you even need to install one, and if so, what guidelines are in effect. In some areas, you need a special permit to install these devices. In my area, where these devices are not yet mandatory, the vacuum breaker is installed in the upright position without the need of any downstream valves, and at a minimum height of 6 in. off the ground.

Vacuum breakers have a frictional loss value (usually several pounds) that is not always listed on the device; this information may be available only through the manufacturer or distributor. If you need to install a vacuum breaker and want an accurate frictional loss figure, you should find out what this value is.

EXPANSION-TANK AND CHECK-VALVE SYSTEMS

When a storage-tank type of water heater makes hot water, the water pressure increases inside the tank. In most installations, the main in-house building supply line connects directly to the cold inlet side of the heater. When the heater cycles and you do not use the hot water that is produced, it backs up out of the heater because it now has more pressure than the cold water feeding it. Under pressure, the hot water is forced back into the supply line that connects to the service line, which in turn connects to the municipal main. If everyone's water heater were cycling hot water back into the main, there is concern that toxic materials could enter a home's freshwater system, be introduced to the water heater and then be reintroduced to the municipal main. Some local codes now require that an expansion tank and check valve (see the photo above right) be installed on storage-tank water-heating systems. Amtrol manufactures an expansion-tank and check-valve system for residential application.

EQUIVALENT LENGTH OF PIPE (IN FEET) FOR VARIOUS GALVANIZED AND COPPER FITTINGS AND VALVES

FITTING DIAMETER (INCHES)	90° ELBOW	45° ELBOW	STANDARD 90° TEE	COUPLING AND STRAIGHT RUN OF TEE	GATE VALVE	GLOBE VALVE	FULL-PORT BALL VALVE
½	1.0	0.6	1.5	0.3	0.2	7.5	NEGLIGIBLE
¾	1.25	0.75	2.0	0.4	0.25	10.0	NEGLIGIBLE
1	1.5	1.0	2.5	0.45	0.3	12.5	NEGLIGIBLE
1¼	2.0	1.2	3.0	0.6	0.4	18.0	NEGLIGIBLE

This chart is adapted from UPC "Sizing Water Systems."

FIXTURE HEIGHT AND HEAD PRESSURE

In designing the supply system, you have to consider the height of each fixture above the water source of supply. Water has weight, and it loses pressure as it is lifted in the piping to reach the various floor heights of the structure. This weight is referred to as head pressure. The more upper-story fixtures you have, the less pressure the water will have in those fixtures, so adequate pipe size is important.

FRICTIONAL LOSS

The longer the pipes and the farther the fixtures from the water supply, the more pressure is required to drive the water forward. Other factors that affect water pressure are the smoothness of the piping material and the mineral content of the water. (Pipes carrying water prone to mineral buildup should be sized larger than they otherwise would be.)

Piping materials differ in their ability to allow water to pass through them smoothly and swiftly. This drag on the flow is usually referred to as frictional loss. For example, for a given number of fixtures, most major codes allow a smaller-diameter pipe of copper than of galvanized steel. This is because copper is smoother than galvanized steel. My major code has frictional-loss charts that categorize pipe materials as fairly smooth, fairly rough or rough.

EQUIVALENT LENGTH OF PIPE

Every change-of-direction fitting (tees, 90° elbows and 45° elbows) adds to the frictional loss of the system. This loss is described in terms of "equivalent length of pipe" (and expressed in feet) and varies with the material and with the type of fitting (for a sample, see the chart above). Look carefully at the numbers in this chart, especially for ½-in. and ¾-in. pipe, which most residential fresh-water distribution systems use. A ¾-in. pipe carries 50% more water than a ½-in. pipe, yet adds a much smaller percentage of frictional loss as expressed in equivalent length. In situations where ½-in. pipe is marginally correct but ¾-in. pipe provides a comfortable margin of safety, using the larger pipe may be worthwhile, especially when you take into account the relative cost of these two sizes of pipe and their fittings.

Valves (see p. 28) are also rated in terms of equivalent length of pipe. As shown in the chart above, a full-port ball valve, for all practical purposes, has almost no frictional loss. A gate valve for ½-in. pipe has an equivalent length value of close to 2½ in. In contrast, a globe valve for ½-in. pipe has an equivalent length value of 7½ ft. That is a terrible restriction and reason enough to choose full-port ball valves or gate valves over globe valves.

Equivalent-length values apply to new, unused fittings and valves. Over years of service, minerals and sediment build up inside the system, slowing the flow of water. If your water supply produces sizable amounts of sediment and buildup, don't skimp on pipe diameter. The small amount of money you save today will be offset by the inconvenience of slow-flowing water in the future.

FRESH-WATER SUPPLY AND DISTRIBUTION

If your house will be served by a city water main and you have neighbors with homes of comparable size and comparable plumbing fixtures, the best, quickest and most dependable way to size your plumbing system is to check the pipe sizes and plans in your neighbors' houses. If your house will be served by a private well, it's still worth checking your neighbors' pump choice and pipe sizing, even though well-pump capacities differ more than city main capacities in the same neighborhood.

If your house differs drastically from your neighbors' houses in height and fixture-unit totals, then you will need to size your system from scratch, as discussed below. Often your local plumbing inspector or building inspector can help you arrive at a satisfactory piping plan. Many plumbing inspectors were once plumbers themselves and are very knowledgeable about system design, especially for their local neighborhood. Judicious consultation with your local authority can do away with a lot of guesswork and misguided solutions, and save you a lot of trouble, especially at inspection time.

Begin sizing the system with a sketch of your home. Note what fixtures are on what floor, sketch the most direct route of supply pipes back to the water source and measure the length of risers and branch lines to the source of supply, either a water meter or a pump house. Also note changes of direction in the piping. This information, along with the total fixture demands for the house, tells you how large a water service you will need. Knowing the distances, both horizontal and vertical, along the planned piping paths will tell you what size the supply piping will need to be.

SIZING COLD-WATER LINES

Now let's work through the sizing of cold-water lines in a hypothetical house, using the requirements of my major code (UPC) in some typical applications, as a lesson on how to approach pipe sizing. Since your major and local codes may differ, you should consult the rules in effect in your area before proceeding with an actual job.

The drawing on p. 46 is a schematic representation of the hypothetical house's fresh-water distribution system. In the drawing, black lines and black capital letters refer to cold-water supply; grey lines and grey lower-case letters refer to hot-water supply. This small two-story house with a basement has one-and-a-half baths (two toilets, two lavatories and a tub/shower), two hose bibbs, a laundry (laundry sink and washing machine) and a kitchen sink — a total of 22 fixture units (see the bottom chart on p. 47 for fixture-unit values for each plumbing fixture). The top chart on p. 47 takes you through the process of calculating the total developed length of the hot and cold supply piping from the meter in the hypothetical house shown in the drawing: about 163 ft. The water supply is a city main with a minimum daily pressure of 75 psi. Because the city main pressure is below 80 psi, a pressure-reducing valve (see p. 42), which would have increased the frictional loss by another 5 psi to 7 psi, is not required.

The first problem is determining the water pressure at the highest fixture in the house, the shower head in the upstairs tub/shower, which is 23 ft. above the supply (in this example, assume rises of 4 ft. from water main to meter, 4 ft. from meter to start of building supply, 9 ft. from first to second floor, and another 6 ft. up to the shower head). Water pressure drops 0.5 psi per ft. of height from the source. so the pressure drop at the shower head is 11.5 psi (23 x 0.5).

The water meter consumes 7 psi of the 75 psi, for a net incoming pressure of 68 psi (75 - 7 = 68). Next we subtract the loss of head pressure (rounded off to 12 psi) from the 68 psi available to the house, to arrive at a minimum house water pressure, or pressure range, of 56 psi.

Now that we know the number of fixture units, the pressure range and the developed length of the supply piping, we can consult the chart on p. 48 to determine the required service-pipe diameter from the meter to the house and the required building supply-pipe diameter. The information in this chart is based on a water velocity of 8 ft. per second. (Average peak consumption for a modest single-family home under "normal" living conditions is about 6 gal. to 8 gal. per minute. Water flowing at a rate of 8 gal. per minute through a ½-in. dia. galvanized-steel pipe travels at a velocity of 8.45 ft. per second.)

SIZING A WATER-SUPPLY SYSTEM

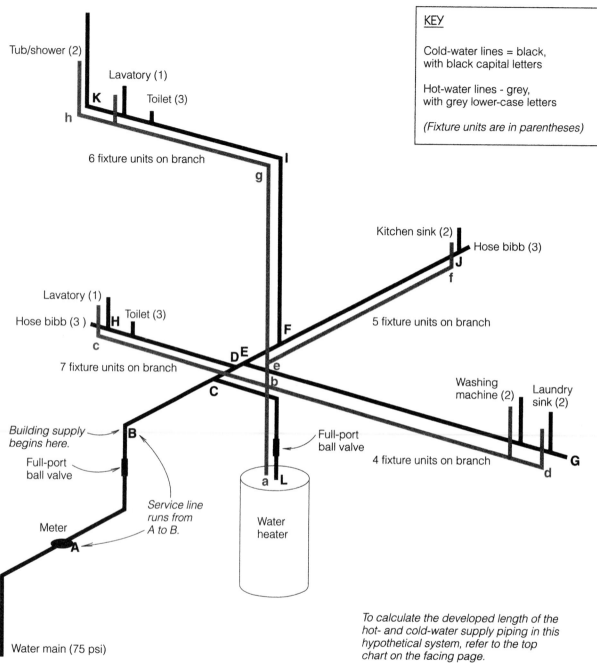

KEY

Cold-water lines = black,
with black capital letters

Hot-water lines - grey,
with grey lower-case letters

(Fixture units are in parentheses)

Tub/shower (2)

Lavatory (1)

Toilet (3)

6 fixture units on branch

Kitchen sink (2)

Hose bibb (3)

5 fixture units on branch

Lavatory (1)

Hose bibb (3)

Toilet (3)

7 fixture units on branch

Washing
machine (2)

Laundry
sink (2)

*Building supply
begins here.*

Full-port
ball valve

Full-port
ball valve

4 fixture units on branch

*Service line
runs from
A to B.*

Meter

Water
heater

*To calculate the developed length of the
hot- and cold-water supply piping in this
hypothetical system, refer to the top
chart on the facing page.*

Water main (75 psi)

CALCULATING DEVELOPED LENGTH FOR THE HYPOTHETICAL WATER-SUPPLY SYSTEM SHOWN ON THE FACING PAGE

COLD WATER SUPPLY (FEET)		HOT WATER SUPPLY (FEET)		TOTALS
B TO F	10	a TO b	4	
F TO J	12	b TO c	10	
D TO H	8	b TO d	16	
E TO G	16	b TO g	9	
C TO L	8	g TO h	10	
F TO I	9	e TO f	12	
I TO K	10			
SUBTOTAL	73		61	134
TOP FLOOR FIXTURE-VALVE RISERS (FEET)				
TUB/SHOWER	3		3	
LAVATORY	1½		1½	
TOILET	½			
SUBTOTAL	5		4½	9½
FIRST FLOOR FIXTURE-VALVE RISERS (FEET)				
LAVATORY	1½		1½	
TOILET	½			
WASHING MACHINE	3½		3½	
LAUNDRY SINK	3		3	
KITCHEN SINK	1½		1½	
SUBTOTAL	10		9½	19½
TOTAL	88		75	163

FIXTURE UNITS AND MINIMUM PIPE DIAMETERS FOR WATER-SUPPLY PIPES

	FIXTURE UNITS	MINIMUM PIPE DIAMETER (INCHES) COLD WATER	HOT WATER
BATHTUB	2	½	½
BIDET	2	½	½
SHOWER	2	½	½
TOILET (FLUSH TANK)	3	⅜	
BAR SINK	1	⅜	⅜
KITCHEN SINK	2	½	½
LAVATORY	1	⅜	⅜
LAUNDRY SINK	2	½	½
HOSE BIBB	3	½ TO ¾	
DISHWASHER	2		⅜ TO ½
WASHING MACHINE (STANDPIPE)	2	½	½

This chart is adapted from UPC "Sizing Water Systems."

WATER METER, SERVICE AND SUPPLY PIPE DIAMETERS

PRESSURE RANGE: 30 TO 45 PSI

METER AND STREET SERVICE (INCHES)	BUILDING SUPPLY AND BRANCHES (INCHES)	MAXIMUM ALLOWABLE LENGTH (FEET)														
		40	60	80	100	150	200	250	300	400	500	600	700	800	900	1000
¾	½**	6	5	4	3*	2*	1*	1*	1*	0*	0*	0*	0*	0*	0*	0*
¾	¾	18	16	14*	12*	9*	6*	5*	5*	4*	4*	3*	2	2	2	1
¾	1	29	25	23	21	17	15	13	12	10	9	7	6	6	6	6
1	1	33	31	27	25	20	17	15	13	12	10	8	6	6	6	6
1	1¼	54	47	42	38	32	28	25	23	19	17	14	12	12	11	11

PRESSURE RANGE: 46 TO 60 PSI

METER AND STREET SERVICE (INCHES)	BUILDING SUPPLY AND BRANCHES (INCHES)	MAXIMUM ALLOWABLE LENGTH (FEET)														
		40	60	80	100	150	200	250	300	400	500	600	700	800	900	1000
¾	½**	9	7*	6*	5*	4*	3*	2*	2*	1*	1*	1*	0*	0*	0*	0*
¾	¾	19	19	19	17	14*	11*	9*	8*	6*	5*	4*	4*	3*	3*	3*
¾	1	33	33	33	33	28	23	21	19	17	14	12	10	9	8	8
1	1	33	33	33	33	30	25	23	20	18	15	12	10	9	8	8
1	1¼	75	75	75	67	52	44	39	36	30	27	22	20	19	17	16

PRESSURE RANGE: OVER 60 PSI

METER AND STREET SERVICE (INCHES)	BUILDING SUPPLY AND BRANCHES (INCHES)	MAXIMUM ALLOWABLE LENGTH (FEET)														
		40	60	80	100	150	200	250	300	400	500	600	700	800	900	1000
¾	½**	11	9	7*	6*	5*	4*	3*	3*	2*	1*	1*	1*	1*	1*	0*
¾	¾	19	19	19	19	17*	13*	11*	10*	8*	7*	6*	6*	5*	4*	4*
¾	1	33	33	33	33	33	30	27	24	21	17	14	13	12	12	11
1	1	33	33	33	33	33	32	29	26	22	18	14	13	12	12	11
1	1¼	75	75	75	75	74	62	53	47	39	31	26	25	23	22	21

* Branch pipes up to 20 ft. developed length (from main to outlet or fixture) may supply maximum of four fixture units for ½-in. I.D. pipe and maximum of 16 fixture units for ¾-in. I.D. pipe.

** Building supply, ¾ in. I.D. minimum.

We first locate the part of the chart that covers our pressure range of 56 psi and find it in the middle section. We next locate the developed length (approximately 163 ft.) under Maximum Allowable Length, which happens to be the sixth column from the left. (If you land close to midway between columns, use the next higher figure.) Next, we read down this column and find that our 22 fixture units falls between the fixture unit totals of 11 and 23. We settle on the higher number, 23, and read across to the left to the headings Meter and Street Service and Building Supply and Branches, which tell us the required service-pipe diameter from the meter to the house (¾ in.) and the required building supply-pipe diameter (1 in.).

In most major codes, ¾-in. dia. service pipe is the smallest allowable size. The 1-in. dia. building supply line was sized on a total demand figure; that is, on the assumption that all of the fixtures would be operated simultaneously — something that almost never happens. Because the pipe sizes in this table are for total demand, we could skimp a bit on the supply and branch lines. However, using the more generous sizing automatically compensates for any future frictional loss in fittings without the need for tedious individual calculations, so we will keep the building supply piping (from B to J in the schematic drawing) a 1-in. pipe.

Now we can size our branch lines just as we did the service and supply lines by using fixture-unit totals and distance to the supply line within the pressure range in the middle chart on the facing page. Branch lines C to L, D to H, E to G, and F to I to K can drop down one size to ¾ in. and still accommodate future buildup/restriction and compensate for frictional loss in the fittings.

SIZING HOT-WATER LINES

To size the hot-water piping, we again consult the drawing on p. 46 and the charts on p. 47 and work through the same steps. We have about 75 ft. of hot-water piping and a fixture-unit total of 10 (two lavatories, a tub/shower, a kitchen sink, a washing machine and a laundry sink). To the pressure range of 56 psi established for the cold water, we add a few psi because the water is heated.

The pressure range puts us on the middle chart on the facing page, and the length figure puts us between 60 ft. and 80 ft., so we use the higher figure of 80 ft. Reading down the scale, we find ourselves between 6 and 19 fixture units, so again we go to the larger figure. Reading across to the left, we find a branch figure of ¾-in. pipe.

Our top-floor hot-water fixtures are a tub/shower unit and a lavatory, which have a fixture-unit total demand of 3. (Most codes assign a value of hot-water demand at 75% of total demand, but because of future buildup/restriction considerations, I like the margin of safety in using full values.) Because all the fixtures that use hot water are unlikely to be operated simultaneously, the branch line from g to h could be reduced to ½-in. pipe. The drop from g to e will be ¾-in. pipe. The 12-ft. run from e to f with only a two fixture-unit kitchen sink at the end can be run in ½-in. pipe. Right below e, we have two branch lines joining at b. The 10-ft. run to c, with only a lavatory at the end, will be easily handled with ½-in. pipe. From b to d we have a 16-ft. run. All these calculations are rough because they don't take into account the change-of-direction fittings (see p. 44), but because the pipes are oversized, the omission doesn't matter. So from b to d the run will be ¾-in. pipe. Our riser a to b to e to g, which is in essence our hot main supply, will also be ¾-in. pipe.

A final word about sizing the water lines, both service and supply. Codes are generally based on minimums, but there is no law that says you must install a minimum system. I don't do minimums, mainly because I rip out so much undersized piping in the service/repair portion of my business that when I am installing water systems, I prefer to do a little more than what is required. I have never found the cost of going up one pipe size to be a hardship on the customer, and I have never regretted the choice.

FIXTURE UNITS AND MINIMUM TRAP AND TRAP-ARM DIAMETERS

	FIXTURE UNITS	MINIMUM TRAP AND TRAP-ARM DIAMETER (INCHES)	MAXIMUM TRAP TO VENT DISTANCES (FEET)
BATHTUB	2	1½	3½
BIDET	1	1½	3½
SHOWER	2	2	5
TOILET	3	3	6
BAR SINK	1	1½	3½
KITCHEN SINK	2	1½	3½
LAVATORY (ONE BOWL)	1	1¼	2½
LAVATORY (TWO BOWLS ON ONE TRAP)	1	1½	3½
LAUNDRY SINK	2	1½	3½
WASHING MACHINE (STANDPIPE)	2	2	5

This chart is adapted from UPC "Drainage Systems."

MAXIMUM UNIT CAPACITIES AND MAXIMUM LENGTH (IN FEET) OF DRAINAGE AND VENT PIPING

PIPE SIZE (INCHES)		1¼	1½	2	2½	3	4
MAXIMUM UNITS FOR DRAINAGE PIPING (EXCLUDING TRAP ARM)	VERTICAL	1	2*	16**	32**	48***	256
	HORIZONTAL	1	1	8**	14**	35	216
MAXIMUM LENGTH FOR DRAINAGE PIPING	VERTICAL	45	65	85	148	212***	300
	HORIZONTAL	UNLIMITED (BASED ON SLOPE OF ¼ IN. PER FOOT)****					
VENT PIPING (HORIZONTAL AND VERTICAL)	MAXIMUM UNITS	1	8	24	48	84	256
	MAXIMUM LENGTHS	45	60	120	180	212	300

* except sinks and urinals.
** except six-unit traps or toilets.
*** only four toilets allowed on any vertical pipe or stack;
    no more than three toilets on any horizontal branch or drain.
**** for slope of ⅛ in. per foot, multiply horizontal fixture units by 0.8.

Note: The diameter of an individual vent shall be not less than 1¼ in. nor less than half the diameter of the drain to which it is connected. Not more than one-third of the total permitted length of any vent may be horizontal. When vents are increased one pipe size for their entire length, the maximum length limitations specified in this chart do not apply.

DRAINS, WASTE AND VENT

Once the fresh-water supply and distribution system has been designed, the DWV system can be planned to accommodate it. Just as with the supply and distribution system, each component of the DWV system is sized according to the number of fixture units of each fixture that it has to service.

SIZING TRAPS AND SELECTING FITTINGS

All plumbing fixtures have a trap (see p. 15), except for the dishwasher, which usually shares the trap of the kitchen sink. The trap arm is the straight fitting that connects the trap to the drain. The plumber installs the trap for every fixture except one: the toilet. Where is the toilet's trap? It is built in; it is part of the casting of the bowl. So any horizontal piping connected to the discharge opening of the toilet is in essence the trap arm. The chart at top lists various fixtures and their fixture units and the minimum diameter trap and trap arm needed to drain each fixture, as well as the maximum trap to vent distances for each diameter trap.

With fittings, size is not an issue since the size is determined by the diameter of the pipe or pipes that they connect. However, fittings come in a vast array of materials and configurations (see pp. 15-21), and selecting the appropriate fitting for DWV piping can be a nightmare for the novice. The wrong choice will impede the flow of water and wastes through the system and drastically impair its effectiveness. When in doubt about what fitting to use in a particular application, consult with your local authority.

SIZING DRAIN, WASTE AND VENT PIPES

Sizing DWV pipes is a lot easier than sizing fresh-water supply and distribution systems. The tables in my code book are so well designed that they eliminate any guesswork. The bottom chart on the facing page, which is adapted from the Uniform Plumbing Code (UPC), is a simple-to-follow guide. As with all codes, the stated pipe sizes are minimums for the given unit loading.

The drawing on p. 53 represents the DWV system for a hypothetical house. In the drawing, the fixture-unit ratings for each fixture are noted, as are the minimum trap diameters. (All plumbing codes stipulate that no drainage pipe should be smaller than any trap arm connected to it. If the drain were narrower than the trap arm, there would be restriction and stoppages, leading to unsanitary conditions.) The DWV system is sized fixture by fixture. Let's begin with an overview of the house's drainage plan.

Most of the fixtures in the hypothetical house drain to the central soil stack, which in turn connects to the building drain; all the vents in this part of the system are back-vented to the vent stack on the top floor. Of the 14 drainage fixture units for this house, 12 drain into the soil stack; only the laundry sink has its own drain.

Sizing the soil stack is the first job. Referring to the bottom chart on the facing page, we see that a 4-in. dia. soil stack, which is a vertical pipe, can drain 256 fixture units for 300 ft. A 3-in. dia. soil stack can drain 48 fixture units for 212 ft. Quite a difference in capacity for one pipe size, wouldn't you say? But in this house, with only 14 fixture units, even 3-in. pipe is more than adequate.

Now let's consider the bathtub on the top floor, which shares its drain line with the kitchen sink. The tub is rated at 2 fixture units; from the top chart on the facing page we see that its minimum trap and trap arm size is 1½ in. Therefore the drain for the tub must be at least 1½-in. dia. (the drain cannot be smaller than the trap and trap arm). Referring again to the bottom chart on the facing page, we see that 1½-in. vertical pipe can drain 2 fixture units, except sinks and urinals. However, this restriction is not always enforced by the local inspector, who may not insist on 2-in. pipe. So our tub drain line down to the top barrel of the combo fitting to the kitchen sink below might be able to be 1½-in. dia. pipe.

Another example of when you need to consult with your local inspector concerns the fitting that joins the kitchen-sink drain line to the vertical drain that begins with the tub above (A the drawing on p. 53). My major code wants horizontal drain lines connecting to vertical drain lines through a 45° branch fitting. Also, my local code wants only a sanitary-tee branch connected to a trap arm. These two requirements are contradictory. My major code wants me to use a combo fitting, but my local authority would be concerned that if a combo were used here, the water seal in the trap could be siphoned off because the drainage would attain a higher velocity passing down the combo's branch. My local authority would probably ask me to use a sanitary tee here because a dry trap is dangerous, and slower drainage on such a short run is inconsequential. Therefore in this drawing, the branch fitting is a sanitary tee.

Below the kitchen sink's sanitary tee, things change. The total fixture-unit load for the drain line serving the tub and kitchen sink is 4 units. Although sinks, like tubs, require a 1½-in. trap and trap-arm diameter, there is also a requirement for a minimum of 2-in. waste. We know that the section serving the kitchen sink must be at least 2-in. pipe. The chart on p. 51 tells us that 2-in. pipe can drain 16 fixture units vertically and 8 fixture units horizontally, so it will be more than adequate for this drain line.

Now let's size the vent pipe for the tub and kitchen sink. From the drawing on p. 53 we can estimate that the total length of both vent lines for the tub and kitchen sink would be about 25 ft. (The bottom chart on the facing page lumps horizontal and vertical vent piping together, so we need not calculate them separately.) We see in the chart that 1½-in. pipe can vent 8 fixture units, with a developed length of 60 ft., which is plenty of capacity for this layout. So we would probably use 1½-in. pipe for these vents.

Next, let's consider the upstairs lavatory and toilet. From the top chart on p. 50 we know that one-bowl lavatories have a fixture-unit value of 1 and a minimum trap and trap-arm diameter of 1¼ in. In the bottom chart on p. 50 we see that 1¼-in. pipe can drain 1 fixture unit and vent 1 fixture unit. Either pipe could be 45 ft. in developed length. The minimum diameter drain for toilets is 3 in. Looking at the bottom chart on p. 50, we see that a 3-in. dia. vertical pipe can drain 48 fixture units and vent 84 fixture units. No problem. So if we wanted to, we could drain and vent the lavatory in 1¼-in. pipe and drain the toilet in 3-in. pipe.

Sewer-pipe and vent-pipe sizing should be considered together, since code calls for as much aggregate cross-sectional area of vent pipe that breaks through the roof as there is of sewer pipe outside the house. (Unlike drains, I don't oversize the vent system because larger vents call for bigger holes through the structure.) This house could have a 3-in. building drain and sewer (the minimum), or it could have a more respectable 4-in. building drain and sewer. A 3-in. drain and sewer require an aggregate vent area of a little more than 7 sq. in.; a 4-in. drain and sewer require an aggregate vent area of about 12½ sq. in. Whichever size we choose, we would continue the vent stack up to the roof, starting at the top barrel of the sanitary tee serving the upper-floor toilet.

There is another vent on the top floor. It rises from the lavatory and toilet on the ground floor and backs into the upstairs lavatory vent before continuing over to the main vent stack. This additional vent means increasing the size of the 1¼-in. vent for the upstairs lavatory where it connects to the dropped branch of the vent tee or combo (B in the drawing on the facing page). This is because the lavatory and toilet on the ground floor, which are back-vented together, have a total fixture unit value of 4. To this we add 1, the fixture-unit value of the upstairs lavatory, because these fixtures all connect and draw their atmospheric pressure through the section of vent pipe between B and

the main vent stack. Referring again to the bottom chart on p. 50, we see that 1½-in. pipe will vent 8 fixture units. However, code requires all toilets to be vented with a minimum of 2-in. pipe, so that entire vent run (from C in the drawing all the way up and over to the main vent stack) needs to be at least 2-in. pipe. If the ground-floor toilet were a tub, all the fixtures could be vented with 1½-in. pipe.

The ground-floor lavatory (1 fixture unit) can drain over to the reducing wye branch (D in the drawing) in 1¼-in. pipe, as seen in the bottom chart on p. 50.

The washing machine in the drawing discharges into a laundry sink, which has a drainage fixture-unit value of 2. This sink is drained on an individual drain and vented on an individual vent. An alternative design would be to drain the washing machine directly into a standpipe (2 fixture units) and drain the sink (2 fixture units) separately. This scenario is lightly penciled in in the drawing on the facing page.

In the top chart on p. 50 we see that the laundry sink requires a 1½-in. dia. minimum trap and trap arm. It could be both drained and vented in 1½-in. pipe. But if we go with the alternative design and install a standpipe for the washer as well as a laundry sink, things would have to change if we put both fixtures on the same drain line. According to the chart, the washing machine needs a 2-in. minimum trap and trap arm. The shared drain then must also be a minimum of 2 in. But, the combined 4 fixture units of these two could still be vented in 1½-in. pipe.

Remember that the pipe sizes we have just arrived at are minimums, not optimums. As I mentioned earlier, it's a false economy to skimp on pipe sizes. For a little more money, you can use the next larger pipe size and be assured that the system will accommodate additional fixtures in the future.

In some areas of the country, the local inspector must approve isometric drawings of a proposed plumbing system before any pipes can be installed. Elsewhere, the system is checked out at the water or air test. Obviously changes at this late stage are inconvenient at best, so it's a good idea to consult with your inspector beforehand if you are unsure of what pipe sizes to use.

SIZING A DWV SYSTEM

*Fixture units are
in parentheses.*

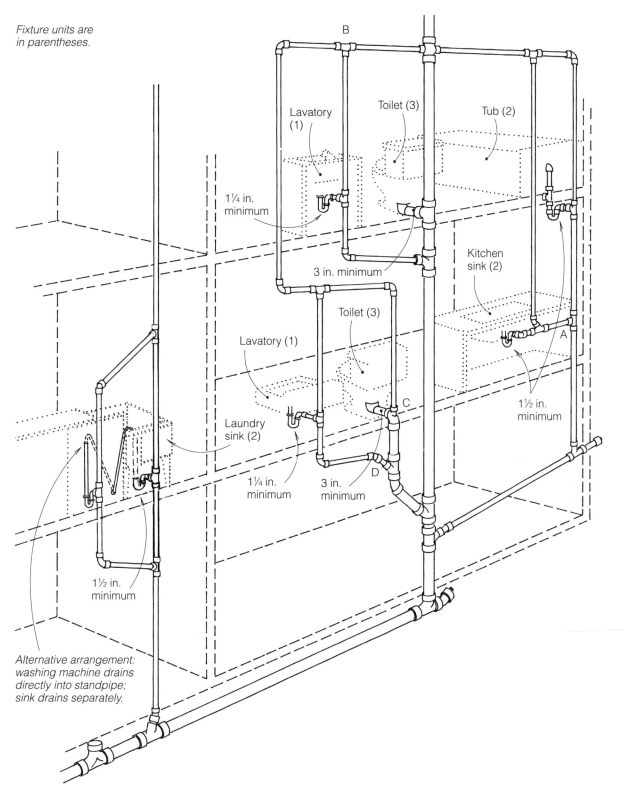

B

Lavatory
(1)

Toilet (3)

Tub (2)

1¼ in.
minimum

Kitchen
sink (2)

3 in. minimum

Toilet (3)

A

Lavatory (1)

Laundry
sink (2)

1½ in.
minimum

C

1¼ in.
minimum

3 in.
minimum

D

1½ in.
minimum

*Alternative arrangement:
washing machine drains
directly into standpipe;
sink drains separately.*

CHAPTER
4

PIPE JOINERY

Pipes can be joined together in various ways. Depending on the material, pipes may be cemented, coupled, threaded or soldered. But whatever the method, the aim is the same: smooth, leak-free joints. I've been doing this sort of work for over 20 years now, and in this chapter I'll explain how to join common DWV and water-supply piping materials — plastic, copper, galvanized steel and iron. Although terracotta is usually no longer a first-choice piping for new systems, you might need to know how to cut it and splice in new fittings for remodel and repair work. It is discussed on pp. 94-95.

JOINING PLASTIC PIPE

Although it can be threaded, plastic piping for residential plumbing is usually joined with fittings and cement. However, you cannot dry-fit the fittings to the pipe and have them in full penetration and in the exact position that they will lie in when the joints are glued. This is because the fittings have a design taper that allows room for glue. It is almost impossible to shove a fitting onto a pipe to full depth and then be able to pull the two apart without using pliers, wood-

en drifts and mallets. Don't use a lubricant (such as petroleum jelly) on the fittings and pipe to facilitate dry-fitting because if a film of grease is left on the pipe or fitting it will interfere with the application of primers and cements. If you do use a lubricant for a particularly thorny area of fitting work, be sure to clean off any remaining residue with a cleaning agent designed for that particular pipe.

Working with plastic pipe involves exposure to noxious fumes from the primer and cement (see p. 23), but if you care enough about your health to take a few simple precautions (long-sleeve shirts, latex or vinyl gloves, circulating fans and possibly a mask or respirator), I feel that you can do this work within acceptable safety margins. Code-approved uses for plastic pipe are discussed on pp. 22-23 and 31-33.

ABS (ACRYLONITRILE BUTADIENE STYRENE)

ABS pipe sections are cut with a special handsaw or a standard carpenter's chopsaw and joined together with fittings using cement (see pp. 59-61). Whatever tool is used, the goal is to end up with a flush-cut end. The outside and inside edge of the pipe should be deburred with a pocketknife so that when the pipe is inserted into the fitting, the bevel on the inside of the hub fills up evenly with cement, making for a smooth joint, and so the outside edge does not chase the cement out of the fitting socket into the pipe.

Not allowing enough space for the excess cement to collect in a cemented plastic pipe/fitting joint could be a problem in a horizontal application if your fixture units are close to the maximum for the diameter of pipe you are using. When you apply liberal amounts of cement to pipe and fitting and join the two together, bottoming them out, the cement can roll up in front of the incoming pipe and puddle in the bottom of the pipe and fitting. In warm weather, instead of flowing to the bottom of the pipe and forming a smooth surface, the cement might set up and form a welt around the inside of the joint, reducing the cross-sectional area of the pipe and slowing down the flow.

Plumbers who are concerned about this situation hold the pipe back ¼ in. from the bottom of the socket. But doing this will slow you down somewhat because you'll have to mark the intended depths on each pipe prior to making up a joint. You can make a gauge for these marks by using a tin can, plastic cup or bowl of a convenient diameter and depth, and trace all the way around the pipe with a silver artist's pencil or white grease pencil. If you have sized your pipe generously, as I recommend in Chapter 3, you needn't be concerned with such a small loss in cross-sectional area. The pipe-in-socket depth issue is important only if you are working to strict minimums. However, because you will be rotating the fitting on the pipe (or vice-versa), you will need to make registration marks on each piece to mark the desired orientation of the fitting.

Sometimes plumbing inspectors will want to make sure that you are using the piping of the proper size and material, so they will ask you to have any pipe identification facing out for easy scrutiny, especially if they are not familiar with your work. In that case, you want to make sure that when you make your registration marks for the fitting angle, the lettering on the pipe will indeed be easily visible to the inspector.

The trickiest part of joining ABS pipe is getting everything in position before the cement sets up. Set-up times vary among brands and with temperature, but are almost always too fast for beginners. You might as well get used to the idea that you are going to waste a little pipe and a few fittings before the job is done. Fortunately the materials don't cost a lot.

You can purchase ABS cement with or without an applicator attached to the can's screw-on cap. Smaller-sized cans have smaller applicators. You will get into trouble trying to assemble 3-in. or 4-in. pipe and fittings using the 4-oz. can of cement. The applicator won't hold enough cement to coat the fitting and the pipe for a proper, slippery and speedy marriage. If you don't get enough friction-reducing cement on the pipe, it won't seat all the way in the fitting, and you will have to throw the whole mess out and start again. The larger the diameter of pipe, the more force it takes to join the fittings. When working with 4-in. ABS pipe, it's a good idea to enlist the aid of another person, who can help push, pull and twist the fitting as needed.

In the 8-oz., 16-oz. and 32-oz. cans, the correspondingly larger fuzz-ball applicator will do the job, but not very neatly. When immersed in the cement, the fluffy dauber loads up fully and you cannot easily and cleanly regulate how much it holds. At some suppliers, you can purchase a wooden-handled bristle brush specially designed for applying ABS cement to 4-in. dia. and larger pipe and fittings; Pasco's version is #4656. The brush gives you much better control because you can get rid of excess cement by dragging the brush against the lip of the can. Also, for applying cement to fittings in the vertical position, the brush greatly reduces the amount of cement that drips on you, the floor or the ground. This brush also works well on smaller-diameter pipe if you take care not to overload it with cement.

When the cement-coated ABS pipe and fitting are pushed together, the pipe will stay fully seated in the fitting; you don't need to hold them together firmly, as with PVC (see pp. 57-58). The cement and fitting manufacturers recommend twisting the pipe one quarter-turn when you have the two pieces joined together to spread the cement around inside the joint,

1. Make matching registration marks with an artist's grease pencil across the pipe and fitting to be joined.

2. Using a pocketknife, debur the outside and inside edge of the pipe.

3. Daub cement around the inside of the fitting and the outside of the pipe.

4. Insert the pipe into the fitting, with registration marks one quarter-turn apart, then twist so the marks align.

5. When the joint is made, wipe any excess cement with a soft rag.

but if you are joining large pipe on a change-of-direction or branch fitting in hot weather and working alone, you might not have enough time to twist the pipe to the exact angle needed before the cements sets up and freezes your fittings in the wrong position. If you have a helper and you have marked the pipe and fitting for the proper angle, you can usually make this quarter-turn twist of the pipe in the fitting and realign the marks before the cement sets up.

After joining pipe and fittings, wipe off drips and excess cement on the outside of the joint with a soft rag as soon as you can, especially if your pipe is foam core and not solid wall (see p. 22). Solvents in the cement actually "melt" the material, creating a solvent weld of the pipe and fitting material. When plumbers use too much cement, the solvent can eat through the thin layers over the core, and the pipe will break.

PVC (POLYVINYL CHLORIDE)

PVC pipe and fittings are joined much like ABS (see above), except that PVC has to be treated with a special purple primer before the cement is applied (see pp. 60-61). Once you cement and make up the joint, however, the primer and cement will almost always hide the mark you have made for aligning the pieces. To get around this problem, some manufacturers of PVC and fittings recommend that you make two marks on the outside of the pipe: the first one at the depth of the fitting socket, and another right behind that. Then you will be able to measure from the hub edge of the fitting to your second mark and know if the pipe is fully seated in the fitting.

While the primer is still wet, coat the pipe and fitting with cement. Both the primer and the cement should be applied to the full depth of the fitting socket and corresponding length on the pipe. You should apply liberal amounts of cement, but not as much in the fitting as on the pipe, to avoid creating a puddle of cement inside the joint. Most excess cement on the pipe will "roll up" and drip off as you join the two parts.

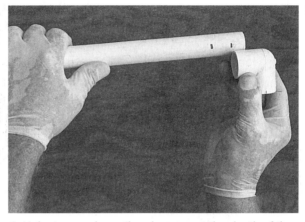

1. Make two marks on the pipe, one at the depth of the fitting socket and the other a known distance behind it.

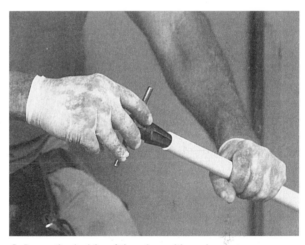

2. Ream the inside of the pipe with a pipe reamer.

3. Bevel the outside edge of the pipe with a knife, then...

4. Apply purple primer to the inside of the fitting.

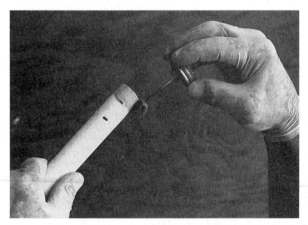

5. Apply purple primer to the outside of the pipe.

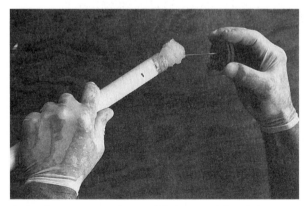

6. Apply cement to pipe and fittings, and then join.

When you assemble the joint, you have to hold the pipe in the fitting for about 30 seconds for small-diameter pipe and about 3 minutes for 4-in. pipe. If you don't, the pipe will back out of the fitting and the joint might fail.

With PVC joinery, temperature will affect your work. In hot weather, if you are too slow, you may not be able to bottom out the pipe in the fitting or the fitting might not be pointing in the direction that you want before it becomes unmovable. Sometimes you can overcome the problems of working in hot weather by storing the pipe and fittings in the shade and by having helpers to apply primer and cement and align fittings. If there is no shade, then you can run a sprinkler on the waiting lengths of pipe and wipe the water off just before cementing. The pipes must be absolutely dry, so use highly absorbent cotton rags for this chore.

In cold weather, you just have to work more slowly and get less done. Hair dryers, paint-stripping guns and torches won't help; they will only cause blisters and bubbles in the cement, which will translate into leaks, not to mention the danger of igniting a can of primer or cement.

Part of the appeal of working with plastic pipe is that little specialized equipment is required. Tools you will need include a pipe saw, a plastic scissors or a miter saw, and, if you are working with PB tubing, a crimper. ABS and PVC pipe and fittings are joined with cements manufactured just for that purpose.

CUTTERS AND CRIMPERS

For cutting both ABS and PVC, I use a Pasco #4333 plastic pipe saw. The tooth set and pitch of this saw are designed to cut plastic pipe without overheating the material and causing the blade to drag. I often use a miter box to ensure straight cuts.

You can buy a wheeled cutting tool designed for plastic pipe that resembles a copper-tubing cutter (see the top photo on p. 65). I don't like these tools because they leave a little lump on the end of the pipe that can cause interference when you are gluing on fittings.

When cutting PVC of diameters up to 2 in., there are quicker methods than the plastic pipe saw. One is the plastic scissors, which resembles a very large gardener's hand-ratcheted pruners. When new and sharp, this tool saves a lot of time. The biggest problem with this tool is finding one that continues to function long enough to justify its cost — about $40. Plastic scissors can

ABS pipe can be trimmed to length with a plastic pipe saw, using a miter box to ensure a square cut.

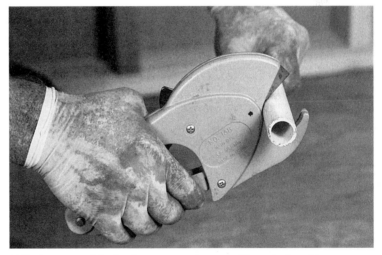

PVC pipe up to 2 in. in diameter can be cut with a plastic scissors.

also be used to cut PB and PE pipe. I don't use the scissors on Schedule 80 pipe or small-diameter Schedule 40 ABS. For those materials the saw works best.

Miter saws can also be used to cut plastic pipe. For a beginner, the finish carpenter's miter saw is a good choice, but it cuts slowly. The power miter saw (chopsaw) with a good-quality, 60-tooth or more finish carbide-tipped blade works quite well on ABS and PVC pipe up to 2 in. in diameter. (To cut 3-in. and 4-in. pipe, you need a very expensive, large blade and saw.) You can now purchase a chopsaw with a slower motor speed that is designed specifically for plastics. The chopsaw is a good choice when you are doing new work and going from room to room on level floors or on a level work surface (plywood next to a trench). In the case of a slab, you might have a helper some distance away on a level surface supplying you with your called-out lengths. Obviously the chopsaw is impractical on repair work under houses or in other confined spaces.

For joining PB pipe you will need a full-circle crimper, which comes with a go/no-go gauge. This gauge, a stamped piece of metal or plastic in the shape of the letter C, is placed around the crimped fitting. If it won't go on, the crimp isn't sufficiently compressed. The crimper has setscrews on the handles that control the degree of crush for the crimp. You might have to adjust the setscrews until your gauge tells you that you are making accurate crimps. There are crimping tools for each size of pipe and crimping tools that will crimp two sizes of pipe. The crimpers that do only one size of pipe are easier and quicker to use, and operate in tighter spots than the double-duty crimpers.

CEMENTS FOR ABS

ABS pipe and fittings use a one-part solvent cement. There is no need for a primer before applying the cement, if the pipe is clean. If you are using up some old pipe that has any oil or stubborn dirt on it, you should clean it thoroughly before doing any cementing. ABS cement is available in different viscosities for various pipe and fitting diameters. For 4-in. pipe, a medium-bodied cement is used. ABS cements work best at temperatures between 40°F and 90°F.

ABS cement is sold at wholesale plumbing suppliers to plumbing contractors in 1-gal. and 5-gal. cans, which can be transferred to 1-qt. cans for more convenient use. It may be cheaper to purchase cement in large containers, but at the point of application, it's better not to use a container larger than 1 qt. One reason is that an opened can lets vapors escape, vapors that are essential for making the cement sufficiently viscous and potent to form a good bond. If you use a 1-gal. can, long before you have used up the material in the can, the cement will probably have become unusable — when this happens, the cement becomes jello-like and won't spread well. A second reason is health. ABS cement is nasty stuff, and the less the vapors are inhaled, the better. A third reason is damage caused by

spills. In the rush to assemble ABS, cement can spill. If it lands on linoleum or other synthetic sheet goods, the damage to the flooring is irreparable.

Two nationally distributed cement lines for both ABS and PVC are Hercules and Oatey. Both of these companies make a cement with a slower cure rate for use in hot, arid climates. Where you will be installing DWV from beginning to end and in some cases for more than one day, you'll want the slowest cure time you can get since the system will have plenty of time (overnight) to set up prior to water testing. For some reason, I find that the cement I buy at the hardware store sets up more slowly than the cement I buy at plumbing-supply outlets. You might want to try the hardware store's cement too, if it is sold in quarts.

The so-called "all-purpose" cements for ABS, PVC and CPVC are clear, heavy-bodied products that allow more time to fidget with ABS pipe and fittings, but your local inspector might not sanction their use, since these products aren't specifically formulated for ABS. But if the local authority has no objections, for new work that will have plenty of time to set before the water test, you could try an all-purpose formula and buy a little extra assembly time.

Cement manufacturers make a "milky-clear" ABS cement that in most cases cures more quickly than the standard black cements. Plumbers who are repairing DWV piping appreciate a fast cure because water can be sent through the pipes almost immediately after the pipe and fitting are joined. Milky-clear cements also dry to a black color.

CEMENTS FOR PVC

PVC joinery requires the application of a primer before the cement. The pipe formula for PVC is a tougher material than ABS. The primer cleans the pipe of grease and grime, but more important, it softens the pipe so that the PVC cement makes a better bond between the pipe and fitting. Most inspectors require the use of a color-tinted primer so they can see right away if a primer was used. Hercules and Oatey make an all-purpose purple primer that is the industry standard for use on PVC. It is the same formula as clear primer, except for the color. Purple primer flows very freely and evaporates miraculously fast, leaving a purple residue (usually a messy one) on both pipe and fitting. The clear primers make for tidier work. If the job is not going to be inspected and the pipe will remain visible, or if you have an exposed system that you want to be as clean as possible, you can probably use a clear primer.

As with other cements, cure times for PVC cement will vary considerably with temperature. General-purpose PVC cement has an average-use temperature range of 40°F to 90°F; Hercules makes a PVC cement that can be used down to -10°F. Low temperatures make the joint assembly vulnerable to failure because installers probably won't wait long enough before bumping or moving the pipe. The cement goes through several physical changes before it arrives at its final hardness. In the can, it starts out like a heavy syrup. Then as the solvent evaporates, the cement resembles the consistency of a gooey mud. Next it becomes rubbery, gelatinous and finally hard. If you move the pipe and/or joint between the mud and the hard stages, you are setting the stage for leaks.

If you are working with 1½-in. to 3-in. PVC pipe in 10°F to 20°F weather, cure time is at least 16 hours. For new construction, a long wait won't be a big issue because you will usually have days to prepare for an inspection where the pipe has to be tested. I have a carpenter friend who works in Alaska on a regular basis at 0°F and he says the other trades are right there with him. However, not many other folks are going to be out working pipe in such cold temperatures except for emergency repairs.

In a more hospitable temperature range of 30°F to 50°F, the cement will set up (not total cure time) within 10 minutes, but allow you at least to handle the cemented pipe and fitting and move on with the job. For water supply and distribution piping, with temperatures close to 70°F or above and with line pressures under 80 psi, I can usually repair a broken line and turn the water back on within an hour using a heavy-duty, fast-set, grey cement.

To join metal and plastic piping, use a threaded male adapter screwed into a female metallic fitting.

JOINING PLASTIC PIPE TO METAL PIPE When plastic joins metal, both materials have to be threaded and joined with adapters. This happens mainly in two instances: in fresh-water piping, where a PVC service pipe meets metal supply pipes, and in DWV, where plastic sink trap arms must join metal piping inside the wall.

Inspectors take a keen interest in the transitional joint between different materials. Most codes do not allow female PVC threaded fittings for water piping because plastic is such a soft material (if a hard, threaded-metal male fitting were used inside the soft, threaded-plastic female, the plastic pipe would stretch and leaks would be inevitable). The code-sanctioned solution is to use a threaded male plastic fitting (male adapter) screwed into a female metallic fitting (female adapter in the case of copper). This joint is assembled using Teflon tape and pipe-joint compound.

Main building service pipes that are done in PVC often need to be joined to copper or galvanized-steel piping at the building. In this situation it's a good idea not to bring the PVC pipe out of the ground at the house. Since plastic pipe is more vulnerable to impact damage than metal pipe, leave all the plastic safely below grade and rise up with either Type L copper or galvanized steel.

Threaded ABS and PVC fittings are tricky to join to threaded metal (galvanized steel and copper) fittings and pipe. In DWV piping, I have found it more difficult to thread a plastic female onto a metal male than a plastic male into a metal female. ABS, PVC, galvanized steel and copper adapters are available with standard pipe threads. But the difference in hardness between the plastic and the metal frequently results in cross-threading (the metal threads

cut into and damage the plastic threads), and once that happens, the plastic piece is no longer any good. You might as well toss it and begin again. When I need to join plastic and metal, I usually have three or four of the plastic fittings that I need before I begin. I very gingerly thread the plastic and metal together, without forcing the fit at all.

The manufacturers of some types of plastic advise against using linseed-oil based pipe-joint compounds (pipe dope) on their plastic threads because it tends to soften them, allowing them to deform easily. Rectorseal #5 is a good-quality, general-purpose pipe-joint sealant that is not linseed-oil based and not a Teflon formula, but you can use it on threaded PVC because PVC is so stable chemically. ABS is not so stable. For ABS, you need a Teflon-formula pipe-joint compound such as #100 Virgin, T Plus 2, or #100-W (made by Rectorseal). Another good Teflon-based formula for ABS is Real-Tuff, made by Hercules.

To make up the joint, I wrap Teflon tape four or five times around the plastic male fitting and apply a conservative amount of pipe dope in the female fitting by dragging the brush applicator from the inside out, across the threads. Once I've got a plastic fitting started properly in the metal fitting, I use only my hands to thread the two together until snug and then apply just an additional half-turn or three-quarters of a turn using slide-jaw pliers.

JOINING PVC TO ABS Should you find yourself in the unlikely situation of needing to join PVC to ABS (the situation might arise in repair work), you might not be able to use the general-purpose clear cement for ABS and PVC because your job is going through fee and design applications and inspection processes. In that case there is a special transition cement you can use. Oatey and Hercules both make transition cement, which is a light fluorescent green color. The Oatey version is #30925; the Hercules version is ABS/PVC Transition. The inspector can tell from a good distance that you used the proper cement for the job. The PVC half of the joint gets the usual purple primer first (see pp. 60-61).

CPVC (CHLORINATED POLYVINYL CHLORIDE)

CPVC feels and cuts exactly like standard Schedule 40 PVC pipe and requires a purple primer specially formulated for CPVC. You must also use cement formulated for CPVC. My local inspectors want me to use orange CPVC cement (it's got a bigger lobby than green CPVC cement). The Oatey version is #31129. The Hercules version is CPVC Orange. If your local authority does not require a colored CPVC cement, you can use a more readily available and less expensive all-purpose clear cement.

PB (POLYBUTYLENE)

Polybutylene is best cut with plastic scissors. It's a somewhat slippery material, but you can mark on the pipe with an ordinary pencil. To cut PB, hold the blades of the scissors at 90° to the pipe so you make a flush cut. The scissors will give you a truer cut if you swing them around the pipe as you squeeze the handles, so they won't flatten the pipe as much while it is being cut.

Brass or copper barbed fittings with a copper crimp ring work the best for joining sections of PB pipe. You slip a copper crimp ring over the pipe and slide it down several inches, then insert the fitting's barb into the pipe until the pipe covers up the last, top barb by ⅛ in. to ¼ in. Slide the crimp ring up until it covers all of the barbs on the fitting, making sure that the copper ring lies flat. Then holding the crimper at 90° to the pipe, compress the ring until the handles contact each other (make sure that the crimpers do not slide the ring out of position). After making the crimp, slip the go/no-go gauge that comes with the tool onto the crimped joint to check tolerances. If the gauge won't slip on, the crimp isn't tight enough.

For PB there are only three types of barbed copper fittings: 90° elbows, 90° branch tees and straight couplings. There are no 45° elbows, and it's not too often that you need one. But should the need arise, you can make one by soldering together a standard copper sweat 45° elbow and two barbed sweat adapters. Do this before you introduce the fitting to the plastic pipe or you'll melt the pipe.

Because PB pipe is flexible, you can make 90° changes of direction without the need for a fitting when space is available. A good rule of thumb for determining the allowable radius of bends is to multiply the outside diameter of the pipe by 12. For example, ½-in. pipe has an outside diameter of ⅝ in, so the minimum radius should be 7½ in. (⅝ x 12). For ¾-in. pipe (outside diameter of ⅞ in.), the minimum radius is 10½ in. If you want to make it easier on yourself, make two circle gauges out of cardboard, one for ½-in. pipe and one for ¾-in. pipe. Then you can easily check the radius of a bend.

When working with PB pipe and tee fittings, avoid acute bends in the pipe attached to the branch of the tee. It is a good practice to come straight off the tee branch with pipe for maybe 4 in. or 5 in. before beginning your radius turn. Also use two pipe supports, one right behind the fitting and one out farther near the beginning of your radius bend, to keep stresses off the tee and pipe.

Because PB pipe is liable to melt when exposed to heat, you have to be careful where and how you install it. Never use PB pipe as the drain for a water heater's temperature and pressure relief valve or for hot-water lines with an instantaneous demand-type water heater. When these heaters malfunction, they often produce steam, which will melt the pipe. PB pipe should be at least 6 in. away from gas heating flues and vents, and at least 12 in. away from recessed lighting fixtures. If you are installing PB pipe in cold-weather areas, use only the new, low-voltage 2-watt to 3-watt heat tapes designed for plastic piping. In the event a PB water line should freeze, use a heat gun, not an open flame, to thaw it out.

JOINING PB TO OTHER MATERIALS For attaching PB pipe to threaded pipe (at the water heater, for example), there is a barbed threaded adapter (both male and female). Use only metal adapters here; plastic fittings can fail because of heat and leakage. For terminating branch lines at sink locations, use a barbed winged female 90° adapter, screwed to wooden blocking inside the wall. Then use brass nipples and standard threaded angle stops of chrome-plated brass or another plated finish on brass.

For valves used with service piping and branches to the main supply, use only full-port ball valves (see p. 28). A barbed ball valve made especially for PB is available in brass and copper. Buy the copper one because it has an integral stop for the handle. You can feel and know when this valve is turned off; you will not turn this valve too far by accident and unknowingly turn it back on. Even better, you can increase your brand selection by using standard FIP ball valves and employing barbed male adapters.

PE (POLYETHYLENE)

PE pipe is cut with plastic scissors and usually joined with nylon or brass barbed insert fittings. These are shoved into the pipe and secured in place with hose clamps. Since this piping will be buried, you should use only hose clamps that are all stainless steel, including the screw itself. If you want an extra margin of assurance, use two hose clamps on each joint.

PE can be buried directly in the ground, but you should not lay it out in a straight line; to allow for thermal expansion, snake the pipe in the trench. You will be safe if you allow 1 in. of expansion per 100 ft. for each 10°F change in temperature.

JOINING PE TO OTHER MATERIALS   It's a good idea to protect PE piping from damage by keeping it safely below grade and coming up out of the ground in copper or another suitable metal pipe material. You can purchase brass insert fittings (barb by FIP or MIP) that will enable you to join PE pipe to threaded pipe and threaded fittings.

JOINING COPPER PIPE

Most people in the trades know that copper pipe is joined by soldering, or "sweating." In soldering, flux is applied to both the pipe and the fitting, the pipe end is pushed into the fitting, and solder is applied at the juncture. When the pipe is heated, solder flows into the joint, where it hardens, forming a seal. The process is called sweating because when the copper pipe gets hot enough for the solder to flow, tiny droplets resembling beads of sweat will appear on the pipe and fitting. Whatever these beads really are (a scientist told me they are drops of hydrochloric acid from the flux), they indicate that the joint has been heated enough to accept the solder.

If you have never soldered before, you should practice on a test pipe and fittings before attempting a real job. If you have access to a workbench with a vise, all the better. Half-inch copper fittings at a big home-improvement store cost very little. All you will need is a scrap piece of pipe maybe 3 ft. long. When the joint cools down, use your tubing cutter and cut it off and do another one. Then, with the aid of a hacksaw, cut your severed pipe and fitting in two lengthwise and see if the solder flowed for the entire depth of the fitting. After a while you will get more comfortable soldering and actually enjoy the process.

A word of warning before you begin. If you are soldering ½-in. copper for a branch line carrying the maximum number of fixture units and you bottom out your copper pipe in the fittings and liberally feed solder to the joints, ridges of solder will probably build up on the inside of the fittings, reducing the effective diameter of the pipe and increasing its frictional loss. However, if you hold back the pipe in the fitting ⅛ in., you leave room for any excess solder and the pipe remains at its designed flow-rate capability. How many plumbers do this? I don't know. I don't have to do it, because my pipes are always oversized.

CUTTING COPPER TUBE

For soldering the ends of the pipe to be joined should be perfectly round. Rigid copper tube is sold in 20-ft. lengths, and often the ends of the pipe are misshapen when you purchase it. Even if the ends look round, I always trim them about ½ in. with the tubing cutter. This does two things. First, the rollers on the lower jaw and the upper cutting wheel of the tubing cutter tend to reshape an oval pipe back into a round pipe. Second, trimming the pipe gets rid of the factory ends, which may be minutely belled or at least slightly ridged, making it difficult or impossible to shove a fitting on the end. A trimmed edge will slide into a fitting more easily.

Most of your cutting will be on fresh-water distribution piping, so your tubing will be Type M, the lightest and thinnest type (see pp. 29-30). Cutting this tubing is a quick and easy operation if you use a good-quality cutter (see the discussion on the facing page). After marking the pipe for cutting (I do this with a mini-hacksaw), you place the cutter on the

Tools for cutting and soldering copper pipe include tubing cutters, fitting brushes, sandcloth, flux, solder and a torch. If you are planning to do flare fittings, you will also need a flaring tool.

TUBING CUTTERS

For water-supply piping, you can get by with only one copper tubing cutter. I recommend the Ridgid #151, which cuts materials from ¼-in. copper refrigeration tube for ice-maker lines to 1½-in. copper pipe and tubular brass. However, the Ridgid #104, which I reverently call "the knuckle buster," is at times required for cutting installed ½-in. and ¾-in. copper tube where there is not enough swing room for the bigger cutter's handle.

For 2-in. DWV copper I would carry at least the Ridgid model #20 or #30. These larger models aren't cheap, but you're throwing money away if you don't buy good quality. (Reed is another good brand.) Cheap tubing cutters tend to thread the pipe instead of tracking in one groove, and they will also slip off the pipe, causing the pipe to deform. If the pipe is out of round, it will be very difficult to get fittings on it and there will be the risk of leaks after soldering.

BRUSHES

Fitting brushes clean the inside of fittings. They are made for each diameter of pipe. The larger sizes (above ¾ in.) get expensive, but

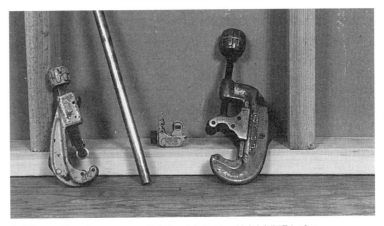

Tubing cutters for copper (left to right): the Ridgid #151, the Ridgid #104 and the Ridgid #30.

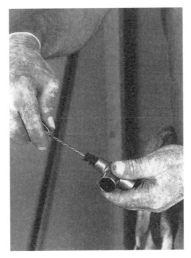

Wire-bristled fitting brushes clean the inside of fittings.

you can use a finger wrapped with sandcloth to do the job instead. Fitting brushes don't last very long before the wire bristles are worn down, but you can usually get through plumbing one house with one brush per pipe size. There are combination fitting/pipe brushes that clean both

the fitting and the outside of the pipe, supposedly eliminating the need for sandcloth. But the brush that cleans the outside of the pipe doesn't last long enough to suit me, and the sandcloth does a better job of cleaning.

Acid brushes are cheap, disposable bristle brushes for applying flux. Even though they cost only about a dime, I keep my "in-use" acid brush in an aluminum cigar humidor when I'm not using it. That way one brush lasts for a long time.

PLUMBER'S SANDCLOTH

Sandcloth is used primarily to polish the end of each piece of pipe before insertion into a brush-cleaned fitting.

Plumber's sandcloth is fabric backed, with a fine grit on its face. It comes in rolls about 1½ in. wide, and of various lengths. Sandcloth is quite economical if you purchase good-quality mate-

Plumber's sandcloth is used to polish the outside of copper pipes before soldering.

rial. I can polish all the pipe and fittings for an entire house with just 1 ft. or less of good-quality cloth. Good-quality sandcloth is water resistant and has good resistance to splitting and tearing. (There is also waterproof sandcloth, but it is too stiff to be used all of the time.) I have found that the dark brown to burgundy cloth has the most water resistance. I have never seen a good blond or light-colored sandcloth, which sells for a lot less money. You can also buy cheaper sandcloth on paper, but I recommend that you avoid it for repair work, because when it gets wet the paper turns to mush. Fabric sandcloth can be dried if it gets wet, and you can continue to use it.

FLUX

Flux cleans the pipe when heated and prevents oxidation at the surface of the metal, allowing the solder to form a strong bond. Flux comes in paste and liquid form. A good flux is critical to successful soldering. There is more than one good brand, but one that is top quality and is sold nationwide is Nokorode. This lead-free paste flux spreads easily, even on wet surfaces. I use this flux on both the fresh-water copper pipes and the larger-diameter DWV copper pipes.

SOLDER

Solder is sold in spools — 1 lb. is the most common size for plumbers, but smaller sizes are available in hardware stores. In the good old days we used to use a 50/50 lead/tin solder for both water-distribution and DWV copper piping. Now new federal regulations regarding lead are in effect and many communities require greatly reduced lead content (5%) or totally lead-free solders for use in copper fresh-water delivery piping. Drain, waste and vent copper piping can still be assembled with leaded solders.

Lead-free solder takes a little getting used to after years of using 50% lead solders. It melts at higher temperatures and cools faster, so you can work more quickly. With 5% lead solder it can be tricky to get leakless joints, but if you do a good job of cleaning the joint and applying

flux things should go smoothly. Lead-free solder is much easier to use than either the old 50/50 leaded or the newer 5% leaded solder. It costs a lot more, but because you can work faster, it more than pays for itself. I use it even on copper DWV piping, where the much cheaper 50/50 leaded solder is still accepted.

TORCHES

Several heat sources can be used for soldering copper pipe and fittings. These include propane, acetylene and MAPP gas.

Propane

The most plebeian tool for heating copper pipe is the small, general-purpose disposable propane cylinder (14.1-oz. capacity; DOT #39 NRC 228-286 M 1003) with a spin-on torch tip (called a secondary-combustion torch). This type is often used to burn old paint off exterior trim. Propane-cylinder capacities are listed by weight, in pounds. A variation on this old design is a squeeze-start, self-igniting torch using the same propane cylinder.

Both of these types have some disadvantages. When you tip the torch end downward, the flame often goes out; also, the flame adjustment does not stay steady when the torch is held at various angles. The tank itself does not work well in confined spaces because it prevents you from getting the flame where you need it. These little torches often do not

Torches and fuel cylinders used in soldering include (left to right): an acetylene swirl-combustion TurboTorch and its fuel cylinders and a Linde secondary-combustion acetylene torch with an A tank. A fire extinguisher is at right in the background.

produce enough heat for good, all-around soldering performance. The usable heat (target heat) in the flame of these propane-fueled secondary-combustion torches is 1100°F, not hot enough for good soldering in all situations.

Another form of the disposable tank type is the swirl-combustion torch, which threads onto the propane tank. This tank gets very hot (the target heat in the flame of a swirl-combustion torch burning propane is 1750°F), and you can tip it to any position with no loss of flame. But the tank makes it hard to get the flame where you need it for all-purpose work.

A big improvement on the tank-threaded torch is the a swirl-combustion torch on the end of a hose, which allows you to keep the tank out of your way and still be able to get in all of the tight spots with the flame. You can even get a belt clip for this type to hold the tank so you can climb ladders and not run out of hose. This torch is good for small repair jobs, but the low-capacity propane tank does not hold enough fuel for you to be using this as your number one rig. If you are constantly buying new tanks, it's an expensive proposition. This little rig cost about $60 the last time I checked.

Acetylene

Older professional plumbers tend to use a heat source that they grew up with — an acetylene torch. Acetylene is very explosive and should be handled with care.

The acetylene cylinders used by plumbers come in two sizes. The one most plumbers use is the A tank, which holds 4 lb. (10 cu. ft.) of gas. (The B tank is much larger and heavier, and is used in industrial applications.) I use an A tank because it is usually large enough for me to get through the soldering for an entire house, yet small enough to drag around for repair work. These cylinders are not throwaways — you have to buy the first one, at a cost of about $40. When empty, tanks can be exchanged at some plumbing suppliers and at welding suppliers.

I use torches that are on the end of a 10-ft. hose so I can work on a ladder or crawl to a location where having the tank next to me would be an inconvenience. Torches that have the blow pipe and regulator attached to the fuel bottle are not as versatile.

The target heat that is generated in the flame of an acetylene secondary-combustion torch is 1400°F. The target heat generated in the flame of an acetylene swirl-combustion torch is 2700°F. I can actually melt the copper pipe and fitting into molten copper with this swirl-combustion torch if no water is present.

I own a secondary-combustion and a swirl-combustion torch (see the photo above left) and each of them does a better job in different applications. The secondary-combustion torch, on acetylene, has a flame that I can make as small or as large as I want, and this flame has a feather that comes to a sharp point. For repair work inside a wall or close to combustible materials, there is less risk of starting a fire if I can train the pointed flame on the pipe and fitting. (I don't care if the repair takes me 10 minutes longer to do.)

The swirl-combustion flame is a squared-off flame that will burn everything that is close to it. However, this flame is much more efficient when heating pipe that has water in it. It turns the water to steam fast enough to keep the cooling water away from the joint so that the solder will flow into the joint.

If you are going to purchase only one good soldering torch, I suggest that you get a plumber's acetylene soldering torch and hook it up to an A tank (the longer the rubber hose, the better). Any commercial welding supply will have one of reputable manufacture. The brand of swirl-combustion torch that I use is TurboTorch.

MAPP gas

Some torches run on MAPP gas, which comes in a throwaway cylinder. MAPP is a mixture of gases, including acetylene and polypropylene. MAPP should be used only with a swirl-combustion torch. It has a target heat of 2400°F, not quite as hot as acetylene itself, but a lot hotter than propane on swirl. These little MAPP cylinders are a potent, but expensive option. Each cylinder costs about $10.

Torch lighters

To light my torch, instead of using the age-old flint, I have a Lightnin' Bug, made by World Wide Welding. It's one of those "magic" crystals that when squeezed, gives off an electric arc. (A flint won't strike when it's wet, and that's a big drawback in plumbing.) Never use a butane cigarette lighter to light your torch. Plumbers and welders have been blowing their innards right our of their chest cavities after those things leak into their clothing and they strike them to re-light a torch.

FIRE-SAFETY EQUIPMENT

Whenever you solder, you should always have a fire extinguisher close at hand. It's safer to solder away from the framing of the house, but sometimes you have to solder copper pipe in place because of logistics; when you do, you need flame shields to protect the surrounding wood. I carry around pieces of sheet metal to slip in, behind and over the pipe and fitting to keep my torch flame away from combustibles. There are also aluminized blankets and high-temperature woven-glass materials you can purchase at plumbing suppliers; these can be pushed, draped or tacked up in areas where stiff sheet metals will not fit to pro-

vide thermal protection to framing and other surfaces. These high-tech materials are not completely flame-proof, but you can reuse them maybe a half-dozen times before you burn holes through them.

TOOLS FOR FLARE FITTINGS

Most of the tools that you'll need to work with flare fittings are simple ones: a tape measure, a 12-in. crescent wrench, an 8-in. crescent wrench, a tubing cutter, a hacksaw with a fine blade, a miter box, a small fine-cut flat file and a pocketknife. The one specialized tool that you will need is the flaring tool. A good one costs around $20, but you can also rent them. Making a flare joint is described on p. 72.

A flaring tool is essential for making flare joints.

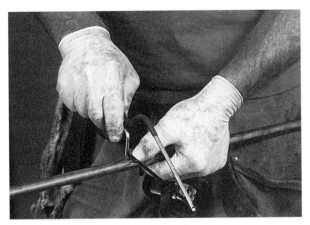

1. Mark the tubing with a mini-hacksaw.

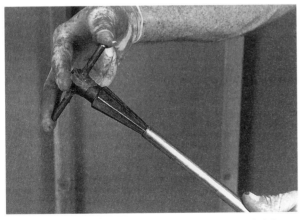

2. Cut the tubing with a tubing cutter.

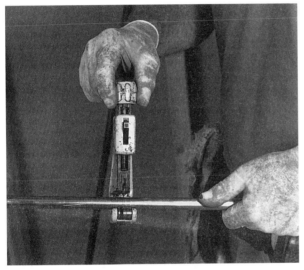

3. Ream the end of the pipe to remove the burr left by the tubing cutter.

pipe and turn the screw handle in until the cutting wheel comes to rest on the cutting mark. Then turn the handle another half-turn and pull it around a full 360°, twice. Tighten the handle another half-turn and go for two complete revolutions in the opposite direction, with the cutting wheel trailing the rollers. Once you see that the cutting wheel is staying in its initial groove, you can continue the rest of the cut in one direction, preferably with the wheel trailing the rollers.

When the cutting wheel of the tubing cutter finally comes through the tube wall, it leaves a tiny ridge sticking out on the inside edge of the tube. This ridge is extremely sharp and can lead to nasty cuts. Remove the ridge by inserting a tapered reamer and returning the pipe to its full bore. This reaming removes the inside burr, but in the process makes the outside edge sharper, so you still have to be careful of cuts. I recommend that you wear a pair of the thinnest leather gloves you can find. Gloves will also give you a better grip than bare hands greased up with paste flux.

SOLDERING RIGID COPPER TUBING

In soldering, cleanliness is the secret to success. Polish every piece of pipe with plumber's sandcloth (see pp. 65-66), down the pipe past the fitting depth, whether or not the pipe looks dirty and scuffed to you. I tear off about a 4-in. long strip of sandcloth from the roll to use on the pipe. I polish the end of the pipe by at least ¼ in. more than the depth of the socket. I loosely drape the sandcloth over the pipe, dividing the cloth in the middle. Then, using my forefinger and the base of my thumb, I squeeze down lightly on the pipe. I rotate the pipe back and forth with my left hand while rotating the sandcloth on the end of the pipe in an opposite back and forth motion (see the photo on p. 66). If you squeeze too hard on the sandcloth, it winds up tight on the pipe and you have to let go and reposition it. Polishing is tricky at first, but after a while you will get the hang of it. Next, use a wire fitting brush to clean the inside of the female fitting (see the bottom photo on p. 65), and put the pipe section and fitting together dry to check for fit.

1. Apply flux to the outside of the pipe.

2. Apply flux to the inside of the fitting.

3. After assembling the joint and heating it with a torch, remove the flame and apply the solder.

APPLYING FLUX AND SOLDER   When pipe and fitting have been cleaned, it is time to don latex gloves and apply the flux. With an acid brush, put a liberal coat of flux on the male pipe and a thinner, but even coat inside the female fitting, making sure to go all the way to the bottom of the socket. Then shove both pieces together. Use the acid brush to remove any excess flux that accumulates at the edge of the fitting after you shove them together. Now roll out about 5 in. of solder from the roll and bend 2 in. back from the end into a 45° or greater angle. In preparation for the next step, put on your safety eyewear and change to leather gloves.

If possible, position your pipe and fitting so that they lie flat, and let them hang off the edge of a bench surface or hold them lightly in the jaws of a vise. Horizontal joints are a little easier to do than vertical, and come off looking better because the solder does not run down the pipe as you feed it into the joint. Capillary action, not gravity, pulls the solder into the joint, whether it is horizontal or vertical. Once the solder becomes molten, gravity pulls a lot more melted solder down from a vertical joint than a horizontal one. It really is harder to keep your vertical joints looking tidy.

Always have pipe inserted into all openings of a fitting before soldering. If you leave one socket open, chances are a portion of it will "tin" (receive a solder buildup) and when it is cooled off, you will not be able to insert a pipe into this opening without re-heating the fitting. If you have good quality-leather gloves, you can gingerly wipe off the excess solder that drips out the bottom of the joints with your index finger.

HEATING THE JOINT   There are several types of soldering torches (see pp. 66-68), but whichever you choose, make sure that it burns cleanly. If you get flux, small solder balls, dirt, wood chips or any foreign matter in the opening of the torch tip, the gas will not fully combust and soot (unburned gas) will be deposited on the joint, where it will interfere with the ability of the solder to enter the joint. You can use flexible pipe cleaners to keep the tip cleaned out. When you buy a new acetylene torch, you will usually get two or maybe even three tips of different sizes. Small tips are intended for small-diameter water pipe; large tips are for large-diameter DWV pipe. The manufacturer will recommend one particular size of torch tip for soldering pipe of certain diameters. My secondary-combustion torch kit came with three different tips. I found that the intermediate tip had no difficulty soldering pipe diameters from ¼ in.to 2 in., even in the heavy Type L and Type K weights.

After selecting a tip, light the torch and adjust the flame. For secondary-combustion torches, the point of the inner blue feather should be clean, not fuzzy. Swirl-combustion torches should have a squared-off flame without a long, thin yellow flame shooting out of the center. Train the tip of the flame onto the fitting socket and start moving it back and forth from the socket to just off the edge of the fitting onto the pipe, and repeat for about 30 seconds, or until you see "sweat beads" on the pipe and fitting. If no beads appear, check the temperature of the pipe by touching the tip of the solder to the joint of fitting and pipe. If the metal is not yet hot enough, the solder will leave no marks or just stick slightly. Keep training the flame slowly, back and forth from the fitting to the pipe.

All of a sudden the little glob of hard solder sitting on the edge of the fitting will turn liquid and be sucked into the joint. At this point, pull the flame back and hold it about 4 in. off the pipe and move the tip of solder around the edge of the fitting, while you feed it into the joint. It is not always necessary to get entirely around the joint. If the pipe cools off too quickly and you cannot get most of the way around, bring the flame back until the solder liquefies again.

Once the gap in the joint is full, droplets of solder will drip out and off the bottom. When that happens, turn the torch off and let the joint cool. Be very careful not to bump, nudge or move the pipe and fitting in any way until the shiny silvery color of the solder turns frosty. (Even the vibration from carpenters driving nails nearby can spoil a joint. I often have had to ask other tradesmen to stop all heavy pounding long enough for a solder joint to cool.) If you disturb the joint while it is cooling, you may cause a leak. If you do disturb it accidentally, reheat the joint until the solder in the joint turns liquid again and once

more remove the flame and let it cool down without interference. If you want to cool a joint more rapidly, you can dip the pipe in a bucket of water once the solder has turned frosty. While the solder is still molten, wipe any excess off with a damp rag.

As you apply the torch, if the pipe and fitting get too hot, the flux may burn away and the solder will roll off the pipe instead of flowing into the joint. If you see a lot of smoke, you are probably burning up the flux. Don't worry about doing anything quickly. You can hold your flame back, farther off the pipe and fitting, to bring the temperature up slowly. Just keep touching the pipe at the fitting until your solder melts and flows.

When soldering vertical joints, always start warming up the lowest joint so that the heat will travel up the pipe and begin heating the next joint up. This way you will conserve energy and time.

SOLDERING COILED COPPER

Coiled copper can be a bit of a challenge to solder because the pipe is not as round as rigid tube. In the process of rolling it up into a coil, the sides of the pipe are "egged out" a tad. A swaging tool can be used to bring the coiled copper back to an acceptable degree of roundness. You insert the tool into coiled copper and then tap on the end with a hammer. The swaging tool is stepped into four or five different diameters for use with various diameters of pipe. I'm always misplacing these handy little tools, so I've perfected my soldering technique to compensate.

When soldering fittings to out-of-round coiled copper, you must deposit a heavy bead of solder around the edge of the joint to fill in the areas where the gap is wider. To do this, prepare the joint as for rigid copper tubing (see the discussion that begins on p. 69), then pull the flame back so that it's just barely hot enough to melt the solder as you continue to build a ring around the edge of the fitting and onto the pipe. As you try to build up this heavy bead around the edge, solder will be dropping off. Don't be alarmed if you have a coin-sized puddle of solder in the dirt below your joint.

FLARE JOINTS

In residential plumbing, soldered joints are not allowed on copper tubing that brings liquid-petroleum (LP) fuel gas from the tank to the house because the gas reacts with lead in the solder, so flare joints must be used instead. The flare for fuel-gas lines is called the straight flare. (The other type of flare fitting, the double upset, is used for very high-pressure lines, as in steam fittings, and is not discussed in this book.) Flare fittings are brass or bronze and work with either type of flare. The flare nuts that secure the tubing to the fittings come in different strengths. With gas, you use extra-heavy flare nuts, which have thicker walls. There are lighter-weight nuts for tubing that carries water, a costly but code-approved application.

The pipe material for LP fuel-gas piping is coiled, soft-temper, Type K copper refrigeration tube. (Refrigeration tube is measured by the O.D.; water tube is measured by the I.D.) Type K, for both types, is the heaviest weight.

Begin by cutting the tubing to rough length (several inches longer than the eventual finish length) with a tubing cutter. You can't make a good flare on a cut end if it is out of round, and the first few inches at either end of a roll of copper tube are at risk of being out of round. Inspect the ends, and if they are misshapen, make a mark with a pencil, scribe or pen far enough back to find good roundness, usually no more than 2 or 3 in.

For the finish cut, use a hacksaw. If you have a small miter box, roll off 1 ft. of tube so you can get it pretty straight. Lay the pipe in the box and cut it on your mark. You have to lay the pipe in the back corner of the miter box and cut only on the push stroke; don't cut by pulling the saw back toward you. File the cut flush and smooth it with a flat file. Then take the sharp edge off the outside corner of the cut, using the file very lightly. If your tubing cutter does not have a reamer blade on the back of it, carefully debur the inside of the cut edge with your pocketknife.

Now slide the flare nut onto the tube, with the belled opening pointing up. Just let go of it for now. On the bar of the flaring tool you'll probably see five holes. Select the proper hole for your tube size. Loosen the wing nuts enough to allow you to insert your tubing in the hole. From below, you want to extend the tubing up beyond the surface of the flaring bar by about $\frac{1}{16}$ in. (You want the side of the flaring

Flare fittings are used on copper tubing that carries LP fuel gas.

bar up with the chamfered holes and tube sizes showing.) Tighten the wing nut closest to the tubing first, and then tighten the second one. Make sure that the wing nuts are really tight, and make sure that the tubing cannot slip down (you cannot pull the tubing out of the flaring bar).

There is a notch in the bottom of the flaring yoke that slides over the flaring bar sideways. Tip up the yoke so that the pointed swivel on the end of the screw is pointing down, over the tube. Turn the screw down with your fingers only, and gently nest it in the tube. When it is centered, apply enough torque with your fingers so it won't move easily. Now turn the handle slowly and evenly. The pointed swivel will form the flare. When you have tightened the handle as far as you can with a reasonable amount of exertion, back it up and remove the yoke. If you were to overtighten the swivel (not very likely when hand-held), you might crack the flare.

Now remove the yoke and flaring bar and slide the nut up so that the flare nests in the top of the cone. The face of the flare mates very accurately to the machined face of the flare fitting. You can now thread the extra-heavy nut onto any flare fitting of the proper diameter. There are flare by flare fittings to join copper to copper, and there are flare by pipe fittings (both male and female) to join copper to threaded pipe.

If you crack a flare now and then, don't get discouraged. The more flares you make, the fewer cracked ones you'll have. That is another reason you should overcut the tube to begin with — you'll have enough room for one or more extra chances to make a perfect one.

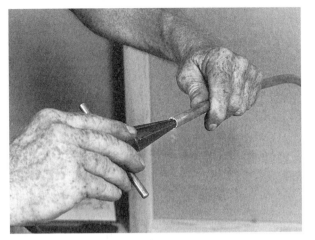

1. Ream the tubing with a pipe reamer.

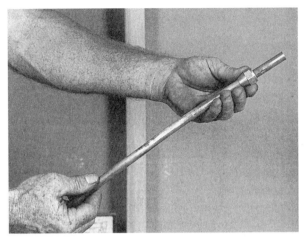

2. Put the flare nut on the tubing.

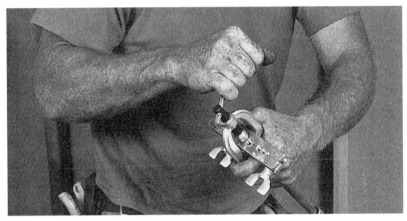

3. With the wing nuts snug, tighten the yoke on the flaring tool.

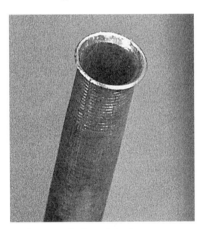

4. A properly flared end.

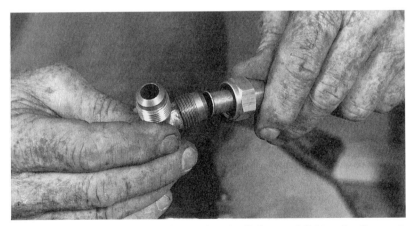

5. Set the flared end of the tubing against the fitting, and tighten the flare nut to secure the joint.

JOINING COPPER IN REPAIR WORK

Sweating new, dry pipe is a dream compared to repairing leaks in existing systems. In new work, the pipes are never filled with water. In repair work, the pipes are always filled with water. Water in the piping complicates the repair process in two ways. First, water left in a pipe will cool the joint sufficiently to prevent the solder from flowing. Second, if you heat the water enough for it to turn to steam (at which point the solder will flow into the joint), if there is no avenue of escape for the pressure built up by the steam it will blow the molten solder out of the joint.

In order to repair existing systems, you need a way to let the water and/or steam escape from the piping while you do your work. In climates where piping is subject to possible freezing (this can occur regardless of insulation), some plumbing inspectors anticipate freeze repairs and call for a continuous downslope on "horizontal" piping so that the building can be drained down at various points through valves soldered to tee fittings installed branch down as an escape for steam and water. In other locales, you might have to provide the escape by other means.

This section of the book is concerned with repair work on copper piping, but much of it applies to remodel work as well, where you have to solder the plumbing for a new room or wing to an already installed system. Remodel plumbing is discussed more fully in Chapter 9.

COMPRESSION FITTINGS AND REPAIR COUPLINGS

Two useful fittings for repairing leaks in copper piping are the compression fitting and the repair coupling. In residential structures, compressions fittings are found in angle stops, dishwasher supply elbows, ice-maker valves, instant hot-water dispensers and water filtering components. In repair work they are used to fix failed solder joints.

The mechanical joints of compression fittings are composed of three parts: the body, the ferrule and the compression nut. These fittings are available in unions, 90° elbows, tees, straight and angled shut-off valves, and hose bibbs. They can be used with both coiled and rigid pipe and tubing with varying degrees of success. Compression fittings usually have flat sides for gripping with the adjustable wrench.

To use these fittings, you cut your pipe or tubing with the tubing cutter. You then slide the compression nut down the pipe or tube an inch or so. Next you slide the ferrule on, and insert the pipe into the fitting or the fitting over the pipe and turn the nuts down, compressing the ferrule.

Compression fittings have really fine threads that cross-thread easily, leading to leaks. The joint can spring a leak if the pipe or fitting is subjected to mild abuse. They also can begin leaking if they are misaligned after assembly. Because of the problems associated with compression fittings, I use them only when I have to. They can be a life-saver in emergencies, however — either late at night or when the leak threatens to disrupt social gatherings. On such occasions, it's often best to install a compression fitting right away, leaving the water supply interrupted, and schedule a more opportune time to drain the line down and make a permanent, soldered repair, as described on pp. 75-77.

The repair coupling lacks an interior stop, which means it can be slid over existing lengths of pipe where space is limited. But this lack of a stop means that the repair coupling takes some extra time to align. To make sure that the coupling is equally spaced over the pipes to be soldered, I nick my two pipes with a scribe or a light stroke of the mini-hacksaw as I hold the coupling under the two sections to be joined. That way, if I have to repolish my pipe with sandcloth, the marks will still be there.

Repair couplings come in three lengths: about 1 in., 6 in. and 12 in. The longer lengths are handy when you need to join pipe sections that are too far separated to use one standard repair coupling and therefore would need two repair couplings and an additional filler section of pipe. These are convenient when you don't want to spend any longer than you have to splicing pipes, or in repairing copper pipe that was split open by freezing, which often causes a rupture that is longer than a standard repair coupling. You can cut the long version with a tubing cutter to any length you need, but you'll have to ream the cut ends before it will slide over the pipe.

FIXING LEAKS Leaks happen for many reasons. Someone may have pounded a nail, screwed a screw or drilled a hole in the wrong place, or sawn too deeply. Many nail holes in copper are caused by carpenters installing baseboard, but the line will often work fine for months to several years before a leak occurs. An errant shovel or pick point might have bashed a hole in a pipe. Improper soldering may have caused a leaky joint, or the pipe may have frozen.

If you have a leak that was caused by a hole from a nail, screw or drill bit, it is usually in a wall, so you have to open up the wall to find the damaged area. If the leak is in a hot-water line, shut off the water to the water heater and drain down the hot water. If the leak is in a cold-water line, shut down the main and then drain the building of cold. With a piece of sandcloth, polish the pipe several inches on each side of the hole. You might have to chisel out some wood from the bottom plate or a stud in order to polish the entire circumference of the pipe. If the hole is not too close to a fitting, you might only need one repair coupling. If it is a small, clean hole, you can use the small repair coupling.

Now hold up the coupling to the pipe. Center the hole in the pipe at the approximate center of the coupling, and then scratch a heavy line on the pipe at each end of the coupling with a nail or scribe. Using a tubing cutter or mini-hacksaw, cut the pipe right on the hole. See if you can push or pull the severed pieces of pipe off center enough to slide the repair coupling onto one or the other legs. If you can, then pull the coupling off, wire-brush it, flux it, flux both pipe legs and slide the coupling down, up or sideways far enough to realign the pipe legs. Then slide the coupling to match the scratch marks, centering it over the hole, and solder your joints (see pp. 69-71).

What if the hole is very close to the edge of a fitting? You'll need a new fitting of the same kind as the old one, and two or more repair couplings to cover each pipe leg cut. What if you cannot push or pull the two legs far enough apart to slide the one repair coupling on? You'll have to make two cuts and use two repair couplings and a new piece to install between them. What if the hole is in a fitting, not in the pipe? Again, you'll need a new fitting of that type and enough repair couplings to join any branches, in the case of a tee or cross (see the photo on p. 76).

Repairing a leaky pipe often means working in cramped quarters. Here a copper pipe is cut in place with the small tubing cutter.

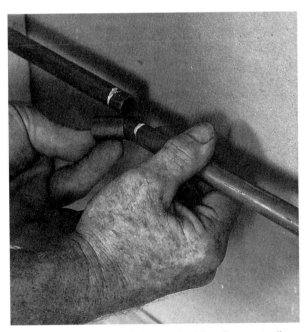

With the cut sections of pipe marked for alignment, slip on a repair coupling and slide it to its intended position against the marks.

Sometimes a nail is driven into a stud or plate and then through the pipe running through that wood member. If you don't want to cut out the wood to reach the hole, yank the nail and cut the pipe at a comfortable distance to either side of the wood member. Pull out the middle section and replace it with a new piece of copper, using two couplings to reconnect it. Then make sure to install a protective nail plate over the pipe (see p. 141).

Resoldering can be a bit frustrating, even with a dry pipe. But if water keeps showing up at the repair location, you won't be able to get the pipe and fitting hot enough for the solder to flow. The solution is to take everything apart and to soak up the water that is dripping out of the pipe ends with bread. I use French bread for several reasons. One reason is storage — a baguette can be months old and bone dry and still be very usable. (Sliced bread is crumbly when it dries out.) Also, the baguette is quite long, so you can introduce as much bread into the pipe as you think will be necessary to absorb the water. The longer the bread, the more time you have for soldering. But, you don't want to introduce any more bread than you have to. If you have low water pressure, you might be waiting hours for the bread to come out through the tub valve and sink faucets after you are finished with the repair.

The bread will absorb only so much water, so speed is of the essence. Plug both sections of pipe as fast as you can, and start soldering. Apply more flux to the pipe, re-assemble the coupling or couplings and any fitting and finish soldering. After soldering, resume water service and open any tub valves first and then faucet valves (after removing the aerators). You want to get as much bread out of the pipe through the valves with the biggest internal passageways. If your repair was on a cold-water line, you don't want to have to take the stems or tops off toilet ballcocks if you don't have to. The bread will pass through faucets easily, but sometimes not too easily through ballcocks.

If you are repairing a vertical pipe and you have standing water in the lower leg of pipe, drop some ¼-in. plastic tubing (ice-maker tubing works fine) down and suck the water out. Wait a few minutes and try again. If you can pick up more water with the tube but it takes a while for it to climb up to the cut, go ahead and fit up the couplings and any fittings needed and then take them down again. Jam a load of bread down the pipe and work quickly to put the pipe back and solder the joints.

If your leak was caused by a saw cut that more or less follows the pipe's length, chances are the gash is longer than a little nail or screw hole, maybe even longer than the small repair coupling. In this case, use a custom cut-to-length portion of one of the long repair couplings that will cover the entire gash and leave enough overlap to approximate the depth of a standard fitting for that diameter of pipe. When you cut your custom-length piece off the long repair coupling, you will need to ream the bore before it will slide over the pipe.

If the leak is the result of freezing, the hole will usually be a 1-in. to 2-in. tear in the pipe. After cutting out the torn section, again cut a custom length off one of the longer couplings, giving yourself at least a ⅝-in. lap on each leg of the pipe.

If the leak was caused by a digging tool, the pipe might also be deformed for several inches around the hole and need a 12-in. repair coupling to span the damaged area. To my knowledge, no one manufactures a 6-in. or 12-in. repair coupling for any copper pipe larger than ¾ in. If you have a copper main that is 1 in. or more in diameter, you'll have to use two standard-length repair couplings for that size pipe.

If the leaky pipe happens to be in a barely accessible crawl space, you won't want to wait around any longer than you have to for the pipe to stop dripping so you can solder it. If water continues to show up at the work area after severing the pipe and jamming in French bread, I look around for a low spot in the same pipe run. If there is none, I try to make one by wedging a piece of wood between the pipe and the floor joist within a few feet of the leak. Then I install a self-piercing saddle valve on each side of the bad fitting and drain the pipes. The water dribbles out the valves, which have been installed hanging down. I then do my soldering repair, shut off the valves and

A long coupling with a self-piercing saddle valve can be used to drain water away from a joint so that it can be repaired.

crawl out to turn the water back on, leaving the valves in place. Perhaps they will be of use the next time another original fitting starts to leak.

When you have to repair a leaking soldered joint, don't try to reuse the fitting that is leaking. You usually can't just reheat a leaky old fitting and then shoot new solder into the joint to stop the leak, though I might try it if I can walk right up to the offending joint and if the pipe polishes up well with sandcloth. But where I have to subject my body to very unpleasant working conditions, I take the surest route: I cut the bad fitting out completely and use a new fitting with one or two new couplings to replace it. Copper pipe that has hung under buildings a long time can be difficult to solder because it has absorbed impurities from the soil. And copper pipe that was overheated when it was soldered originally is very difficult to resolder.

JOINING STEEL AND NO-HUB IRON PIPE

In this section, I'll be discussing how to cut and join galvanized steel, "black" (unplated) steel, protectively coated steel pipe, and no-hub iron. The steel pipes can be threaded to interlock with a mating threaded pipe. Unthreaded no-hub pipe is joined with no-hub couplings. Steel pipe can be cut with a ferrous pipe cutter (big brother to the tubing cutter for copper), with a reciprocating saw or hacksaw, or by machine, with a tool called a "mule" (see p. 80). The threads can be cut manually or with a power threader. With protectively coated steel pipe, the only difference is that the tough plastic coating must be cut back before you can thread it. Iron pipe is cut with a soil-pipe cutter, nicknamed the snap cutter (see pp. 88-89). You can also use an abrasive cutoff saw with a 16-in. to 18-in. dia. blade, but this costly tool can't be found at rental yards, so I won't discuss it in this book.

THREADED STEEL PIPE

On new, full lengths of steel pipe, you will find threads already cut on each end. Traditionally, one end will have what is called a merchant's coupling and the opposite end will have a vinyl cap. The merchant's coupling (see the photo on p. 166) is a thin-walled steel coupling — smart plumbers throw it away because it stretches too easily and will be a source of leaks. But like the vinyl cap on the other end, it does do a good job of protecting the precut threads.

The factory-cut threads are fine to use as they are, if you are joining full lengths of pipe. Often, though, you will have to cut the pipe to length yourself and thread it to mate with various fittings.

CUTTING STEEL PIPE   When cutting pipe with a mule (see p. 80), you slide the pipe down the tube and leave more length than needed sticking out of the locking jaws, with both the cutting arm and threading arm thrown back, out of the way. You mark the desired cut on the pipe, then slide the pipe close to the point where the cutter will lie when it is lowered into position. Now you lower the cutter, with its jaws opened far enough so that it falls onto the pipe gently without contacting the cutting wheel. If you

With the pipe held securely in a chain vise, the ferrous pipe cutter begins its cut.

need to make a fine adjustment, you can either slide the pipe so that the cutting wheel is right over the mark or slide the arm with the cutting tool if it is movable. Now use the handwheel to set the locking jaws, making a final check for the mark. Switch on the machine, slowly close the lower jaw of the pipe cutter and begin cutting the pipe. The pipe cutter remains still and the pipe rotates inside of its rollers and cutting wheel. Slowly close the lower, movable jaw of the cutting tool until the pipe is severed.

When cutting pipe in a truck-mounted vise or a freestanding tripod vise, there is enough room to use the ferrous pipe cutter, a rather large tool (see the facing page). Use this cutter pretty much as you would the tubing cutter for copper (see the discussion beginning on p. 64), but because the ferrous pipe cutter has wider jaws and rollers, you can continue cutting in one direction without having to back up.

The cutting wheel of the pipe cutter will leave a tiny ridge or bulge on the very outside edge of the pipe, which is a big help in getting the pipe threader to start chewing its way down the pipe and form the threads. The four cutting dies in the threading head are designed to produce the taper of standard pipe threads. That little bulge catches quickly on the knife portion of the tapered dies, immediately pulling

Specialized tools necessary to cut and thread steel piping are a vise, a ferrous pipe cutter, a reamer and a threader, either a hand tool (ratchet threader) or a machine (mule or tripod).

PIPE VISE

To cut and thread pipe, you will need a pipe vise to hold your work. Home workshop vises and mechanic's vises sometimes combine pipe jaws in their design, but these usually are a poor substitute for a real pipe vise. Pipe vises are designed to open and close quickly without interfering with the loading and unloading of the pipe. I have a Ridgid chain vise (a #BC 510, for ⅛-in. to 5-in. pipe) bolted to the rear bumper of my truck. The next time you see a plumber's truck, take a good look at the design of the vise. The most common is a vertical screw type, with jaws arced and cut specifically for pipe.

Tools for working with steel pipe include (left to right): a reamer, a ferrous pipe cutter, dies of various sizes and ratchet threaders.

A chain vise bolted to the bumper of a truck will hold steel pipe for cutting or threading by hand.

A pipe-cutting work station called a tripod is favored by plumbers who specialize in pipe threading (poor devils!). Tripods have a large, flat work surface and a built-in or add-on vise. There is a place to hang the pipe-cutting tools, as well as a recirculating oil can and shavings trap. The tripod also has folding legs so it can be carried indoors and set up in whatever stand-up space you are working in.

Tripods cost a lot, and only those crazy enough to seek out this tedious, greasy work usually buy them. Even a modest tripod that must be bolted to a suitable, stable surface (such as a truck body) is rather expensive. If you are going to do your first major threaded-pipe job, you might want to rent a tripod with a vise before purchasing one.

FERROUS PIPE CUTTER

The ferrous pipe cutter resembles a gigantic copper tubing cutter, with rollers in the lower jaw and a cutting wheel on the upper jaw. This tool (mine is a Ridgid #2A) is a necessity if you want to hand-cut pipe in preparation for threading. The larger the pipe diameter, the harder it will be to work with. Residential plumbing rarely requires cutting and threading pipes larger than 2 in. Cutting a lot of 2-in. dia. pipe by hand is beyond the physical capabilities of most adults. Any pipes larger than 2 in. are best subbed out to a specialist.

PIPE REAMERS

There are several designs of pipe reamers. The most common is a beveled cutter that fits the die head of the ratchet threader. You have to change your die for the reamer after each cut. Three or four pulls on the ratcheting handle will usually remove the offending ridge. A brand-new reamer can do the same job in fewer pulls.

Changing from die to reamer and back is a tedious operation, so you might want to use a hand reamer instead. I have a General brand (#135) that I use for ¼-in. to 2-in. dia. threaded pipe. I mount this reamer in a brace to rotate it quickly.

RATCHET THREADERS

I have two sets of ratchet threaders (heads with handles) and dies. One set is smaller than the other. When threading pipe at the vise on the bumper of my truck, I use the larger head, which has a longer handle. The extra length provides greater leverage and ease when pulling. When I have cut a hole in a wall under a sink or behind a tub wall or in a tight crawl space and I need to thread a pipe, I use the smaller set because it can get it in places where the larger set would not fit. The small set is a Ridgid #00R, with ½-in., ¾-in. and 1-in. dies. The large set is a Ridgid #12R, with ½-in., ¾-in., 1-in., 1¼-in., 1½-in., and 2-in. dies. (You can purchase a ⅜-in. die for either set, but that's a size I never use. Plumbers who work with threaded fuel-oil lines commonly use ⅜-in. tools.)

MULES AND TRIPODS

Mules and tripods are the two types of powered threading equipment. The mule, the heavier of the two, is used on jobs where a lot of pipe needs to be threaded. It cannot be picked up by one man. Even two men have a difficult time moving it. Most of these units stay in the truck, mounted to the tailgate, if the truck can get close to the work area.

The mule resembles a small steamer trunk with a big, sliced, summer-squash-shaped chuck on one end. It has forward and reverse gears. The pipe to be cut is slid from the opposite end of the chuck into a tube running down its length. A large-diameter handwheel on the chuck spins in two directions to open and close the gripping jaws. The handwheel has some built-in backlash that is designed to allow the operator to turn the wheel in a slamming movement to set the locking jaws tightly.

In front of these locking jaws, the threading die is attached to an arm that is hinged on the back side. This arm also slides down two parallel shafts that lie parallel to the pipe tube. The threading die has a handle that spreads the thread-cutting dies apart.

Mules have a built-in automatic oiling mechanism that keeps the pipe well lubricated as it is being threaded. Behind the cutting dies on the mule's arm is another arm; it is also hinged on

the back side and hosts a pipe cutter, which looks and operates just like a hand-held pipe cutter.

The tripod, a much lighter machine than the mule, is designed for true portability. It can be carried around by one stout fellow (or two not-so-stout fellows). One of the legs is actually the motorized rotator. You usually need a separate pipe stand, equipped with a roller, to offset the weight of the pipe hanging behind the tripod. The tripod has a smaller locking jaw mechanism than the mule, and all of the cutting tools are standard, hand-held equipment. There is a "rest," upon which you lay the long handle of the threading head so that you are not spun around like an airplane propeller trying to overcome the rotational forces of the rotator. This tool is geared very low.

There is no automatic oiling with the tripod, You usually place a portable oil pot on the ground under the end of pipe being threaded. The oil pot has a screen in it that traps the falling chips of pipe material as they peel out of the threading tool. On the side of the pot is attached a long hose with a hand-held, hand-operated oiler that sucks up oil from the pot and pumps it onto the pipe as it is being threaded. As the oil drips off the pipe it goes through the screen and into the pot again.

Before a protectively coated pipe can be cut or threaded, the plastic coating must be removed.

the tooth-cutting portion of the dies onto the pipe, more or less self-feeding the die head as it moves down the pipe.

Pipe cut with a hacksaw or reciprocating saw has no little ridge or bulge. When the squared edge of this pipe encounters the knives of the hand threader, it is chamfered for several complete turns or more before being drawn into the teeth and being self-fed from there on — if you are lucky. Often you have to shove on the die head firmly from behind as you pull the handle to get the threading process started, even with brand-new dies. This is no big deal if you have lots of maneuvering room about the pipe. But when you are lying on your back in a 2-ft. crawl space, reaching around a pier, it can be very frustrating. So I use the pipe cutter whenever I can.

Ferrous pipe cutters can be used on black and galvanized steel in the pipe vise, but not with protectively coated (machine-wrapped) steel pipe. If you try cutting this pipe with the pipe cutter, the asphalt layer between the coating and the steel will act as a cushion and cause a "wave" to form in the coating in front of the pipe cutter's rollers. The tension on the back side of the wave will cause the outer plastic coat-

ing to tear and and jam in the rollers. Then you will have to back the cutter off and clean out the jammed plastic. It's tempting to use the reciprocating saw instead of the pipe cutter, but if you do you won't have the benefit of the bulge when it comes time to thread.

A better way to proceed is first to remove enough plastic for threading — 4 in. to 6 in. — so that the ferrous pipe cutter can be used. It takes two complete cuts with a knife around the pipe and then one down the pipe between the two radial cuts to remove the plastic. Also remove the protective plastic coating where the pipe vise will grip the pipe when it is threaded (see the photo above).

After cutting your pipe, you should ream the inside edge of the pipe to eliminate the ridge left by the pipe cutter or saw. If not removed, this ridge will reduce the flow and increase noise in the pipe. With the ridge gone, the pipe is ready to be threaded.

1. When threading by machine or by hand (as shown here), the liberal application of oil is essential for a smooth cut.

2. Metal shavings that clog the hole of the die can be removed with a needle-nose pliers.

THREADING   Pipes may be threaded using powered threading equipment or hand threaders. With either method it is critical to apply oil to the dies about every second revolution of the die head to lessen the friction and heat. Thread-cutting oil is sulfured, so you cannot substitute any type of general-purpose automotive oil. There are two formulas: dark and light. Dark oil protects your dies much better than light, for both power and hand threading. Light oil is just what its name implies — lightweight. I have recently discovered that virgin olive oil works better than thread-cutting oil.

Wear gloves when threading pipe to protect your hands from the sharp edges and metal shards. If you rent powered threading equipment, I strongly recommend that you use protective eyewear. If you have also cut your pipe using the mule (see p. 80), you can just leave the machine running in the proper rotation for threading, lift the cutting tool up and back and lower the thread-cutting arm down with dies closed, ready for cutting. Then turn on the circulating pump for the lubricating oil and train the built-in flexible spout onto the end of the pipe to be threaded. Slide the dies forward on their parallel rails. When they make contact with the ends of the rotating pipe, they are dragged onto the pipe and begin cutting threads.

With the mule, when the proper number of threads has been cut (the pipe is now flush with the visible opening in the die head) you either quickly spread open the die head with a lever on top and slide the die head back off the pipe or switch off the machine momentarily and then throw it into reverse

3. The finished threads.

4. A final reaming removes the burr left by the ferrous pipe cutter.

and the die arm backs itself off the pipe. Experienced operators in a hurry often split the dies manually (the first alternative), which is a lot faster.

Next, you have to ream the pipe to remove the burr left by the cutter. Pick up the pipe reamer, which is often a separate tool on a long handle, hold it in the opening of the pipe and ream the pipe as it continues rotating. (Some machines have another hinged arm that holds the reamer, or the threading die may be removable and the reamer installed in its place.) After reaming the pipe, turn off the oil pump and the machine, reverse the handwheel to open the locking jaws and remove the pipe from the tube. Turn it around and thread its other end.

Hand threading also creates shavings that can bind up in the die head and cause you to break or gall the new threads or break a tooth on one of the dies. You cannot mistake these shavings — they look like wound-up coarse steel wire. If you stop threading and then back up a half-turn on the die head, you will break off the shavings from the pipe; then you can use the acid brush to sweep them out through the side openings of the die head. If the acid brush is not stiff enough to do the job, you can use a little hooked scribe or a needle-nose pliers to yank them out.

Look at the number of factory threads on the length of ½-in., ¾-in. and 1-in. pipe that you have. It will probably be between 10 and 12. When you are hand-threading your own pipe from fresh-cut ends and see your pipe flush with the outside of the dies in your hand-operated die head, you will have eight or nine threads. Crank on until you have one or two exposed threads sticking out past the dies, and then call it quits. This will give you 10 or 12 threads. Back the dies all the way off and then ream the pipe end, if you did not do so after cutting it to length. I mount my tapered reamer on an old trusty brace. This saves a lot of time because I don't have to switch from dies to reamer and back to dies on the ratcheting head.

ASSEMBLING THREADED PIPE When installing a threaded-pipe system, it's much easier to start at one end and work in one direction. There is one exception. When running a water line in the ground or suspended from the first-floor joists over a crawl space, you can work from opposing directions and join your sections with a union, a three-part fitting that may or may not be sanctioned by code in your area (check with your local authority). Measuring and fitting unions takes more time, of course, but if you are not a seasoned pro at running threaded pipe, completing the runs that are most evident can help you to visualize the finished system. Unions should really not be installed inside a wall; and, needless to say, the fewer the unions, the fewer the potential leaks in the future.

Assembling threaded pipe is not a complex operation. You wrap the male threads on the pipe with four layers of Teflon tape and apply a thin coat of pipe-joint compound to the female threads in the fitting (see the sidebar on the facing page), then thread the two together. When assembling straight runs of threaded pipe up to a few feet long, it makes little difference whether you thread the male pipe end into a stationary female fitting or turn a female fitting onto a stationary male pipe. With long, heavy sections of pipe (6 ft. or more) and straight female couplings or tees, it is much easier to thread the coupling onto the pipe. It is hard to judge alignment while you are rotating a long piece of pipe and at the same time trying to level or plumb it. The longer the pipe, the more difficult it is to align. When making changes in direction with tee branches or elbows and I find I can't swing the long piece of pipe at this angle, I thread a short nipple into the fitting and then thread a coupling onto the long joining pipe. Then I am once again in the better position: the rotating female meshing with the stationary male.

Most threaded fittings have a mold line, which can serve as an alignment aid. By this I mean that you can look at your pipe from the side and try visually to put the mold line in the same plane as the pipe. However, the mold line is not always a true measure; occasionally you will find that the tapped opening is at an angle to the mold line. If attempting a marriage of pipe and fitting using the mold line doesn't work, you must ignore it and try slight changes in angle. Also, the faces of the fitting may not be plumb, and the ensuing visual distraction may throw you off.

Just how tightly should you assemble threaded parts? There is no pat answer. As long as your system does not leak at its test and operating pressure, it was assembled with sufficient torque. You can crack iron and steel fittings by overtightening them. How round the pipe and fitting are and how accurately they were threaded are factors in the final fit. With cheap imported fittings, I must spend more time with my pipe taps (see p. 7), cleaning out and truing up threads (the fittings may have been dropped at the factory). If the very first thread of the fitting or pipe is impaired, it can be difficult to impossible to mate it, without making repairs. When assembling good-quality ½-in. to 1-in. steel pipe and fittings with 18-in. pipe wrenches, I torque the mating parts until I feel a slight twinge in my elbows. At this point I will usually have four threads left exposed. If I'm installing a coupling, I'm done.

For change-of-direction fittings, I then continue to bring the fitting around until it points in the correct direction. If you exceed this final position and another complete revolution will require too much effort or possibly crack the fitting, you have a problem. Don't just back the fitting up — this is a risky move that invites leaks. You are better off unthreading the fitting from the pipe and starting over, even with a new fitting. The process of threading applies a certain amount of "crush" to the threads on the pipe and fitting. Think of this crush as an altering force. If you reuse the original altered fitting and stop short of your first miscalculated position, the engaged threads may not have enough friction to form a leak-free joint. However, if you have wrapped the threads of the male pipe with Teflon tape and applied pipe-joint compound to the female threads, you can usually back up the fitting as much as a quarter-turn without producing a leak. As long as you don't use excessive amounts of pipe-joint compound, you will not cause any problems. You can apply gut-busting torque with oversize wrenches using just tape or just dope on steel pipe and fittings and still have a leak. Use both the tape and dope and you will have far fewer leaks.

ACHIEVING WATERTIGHT THREADED JOINTS

Whatever the piping material, when you work with threaded pipe you must apply a sealing material to the threads to achieve a permanent leakproof joint. The two sealants I rely on in my work are Teflon tape and pipe-joint compound.

TEFLON TAPE

Teflon tape is used on male pipe threads as a seal against water or fuel-gas pressure leaks. High-quality Teflon tape is of uniform thickness, has good stretchability, does not fray when you tear it, and comes off the spool without sliding off at the edge when under tension. Some companies manufacturing quality tape color their product to differentiate it from cheap brands. One excellent brand sold nationwide is Pink Plumber's Tape, manufactured by the Mill Rose Company.

Teflon tape comes in several widths; the most common are ½ in. and ¾ in. For 90% of my work, ½-in. tape is adequate. The ¾-in. tape is useful for 3-in. and 4-in. dia. pipe threads. When buying tape, look on the spool to see how much tape you are getting for your money. A common length is 260 in., but some stores sell spools with as little as 100 in. for a premium price. When applying the tape, wrap tightly in a clockwise direction and cover all the threads.

PIPE-JOINT COMPOUND

Pipe-joint compound, or pipe dope, as it is commonly called, is used as a seal on female pipe threads (and occasionally on male threads, too). My first choice for pipe dope is Rectorseal #5, which is good for all pipe materials except ABS. (For ABS I use Rectorseal #100 Virgin or Hercules Real-Tuff.)

Rectorseal's #5 is that company's most widely distributed and economically priced compound for the requirements of residential plumbing. It has the runniest consistency of all the choices, which can be an advantage, but it also has the most offensive aroma of all the brands. Rectorseal and other companies also manufacture a pipe-joint compound for assembly in extreme cold weather. The #100 Virgin and the Real-Tuff are stiffer and require more work (causing more mess) to force them into confined areas. Most pipe-joint compounds come in cans that have a brush attached to the screw-on cap (though a squeeze tube would be more practical).

In commercial plumbing, which might carry a variety of gases including oxygen, purified air, process waters and any number of chemicals, pipe-joint compounds have to be compatible with the substance being piped. In residential construction, however, threaded pipe is only for transmitting fresh water and fuel gas, and the same pipe-joint compound can be used for both. I am familiar with about a half-dozen brands of compounds suitable for this application, and each has its pros and cons. One of the brands that I use often is Rectorseal #5, because I can apply it to threaded PVC fittings for water, to galvanized pipe and fittings for water, and to both galvanized and black steel for natural and LP fuel gases.

A lot of joint compounds have a set time. This is the length of time before the compound becomes hard or stiff or something in between. But if you apply Teflon tape to the male threads, the hardening of the compound will not have the effect of cementing or gluing the two pieces together. (Some compounds actually become a cement when they set, and it can be almost impossible to undo a joint that has not been taped once it has set up.) If you take care and slowly but very firmly snug up the joint, you will apply enough compression to ensure that the mated pieces do not loosen up. It is also important that you do a good job of hanging and/or supporting the pipe as you progress (see pp. 34-38).

If you have doubts about any particular joints as you assemble your threaded-steel pipe system, it makes sense to cap or plug off the system at that point and pressure test (see pp. 128-130 and p. 158). Otherwise you might have to disassemble more joints later to correct for leaks. On a gas line, you must use air for the test (see pp. 175-176). On a galvanized-steel water line in a remodel, where draining water out of a partially completed line could cause unacceptable damage, air can be used also. However, it is more difficult to contain high air pressure than high water pressure.

When you have finished assembling a gas or water piping system, especially a threaded piping system, you should purge the lines before putting it to use. In the case of a water line, small fragments of metal threads, bits of Teflon tape and excess pipe-joint compound will ride down the inside of the pipe and lodge in your faucets and other valves. Or you may have a potential blockage from dirt, wood chips, dead bugs or other everyday materials. Gas lines are less prone to blockage from thread chips because the gas-line pressures are only a fraction of water-line pressures and the heavy chips will not migrate so easily, but dust, dirt and small cuttings of Teflon tape will go down the pipe and lodge in sensitive regulators and diaphragms. It easy to purge the water line using the building's supply source or a neighbor's garden hose. The gas line is a different matter. If you can attach the line from your nail-gun compressor via threaded adapters, it will send sufficient compressed air to purge an open-ended line. You could also rent a conventional, portable compressor for this task.

NO-HUB IRON

In the old days, assembling iron DWV piping entailed a lot of work and specialized equipment. You had to have a lead pot, a portable gas burner and fuel tank, oakum (asphalted hemp), lead ingots, and some specialized hand tools. The pipe and fittings were referred to as bell and spigot: One end was smooth (the spigot) and the other end was belled. When pipe and fittings were arranged together, the gap between the male spigot and the female bell was filled in with oakum and hot lead. Many, many buildings have drainage systems of this design, which were "corked" (assembled) 50 or more years ago and are still in functioning order.

In some parts of the country, you still might see a plumber assembling a drainage system using bell and spigot pipe. For the most part, however, no-hub (hubless) iron pipe and fittings have replaced bell and spigot in modern residential DWV plumbing. No-hub pipes are cut to length and joined together using a rubber coupling with an outer stainless-steel wrapping that is cinched onto the pipe and fitting with worm-drive hose clamps. It's a magnificent improvement over corking bell and spigot.

When working with no-hub pipe, you have to check carefully for defects such as cracks. Cracks mean leaks. No-hub pipe is very strong stuff until it is dropped on rocks or concrete; then it can become shards. A piece of pipe that is dropped when it's being loaded for shipment may not shatter, but it will crack. Cracked pipes and fittings have no ring when tapped lightly with any metal object. Instead of a ring, you'll hear a clack-like noise. Get in the habit of lightly tapping the pipe and fitting with the handle of your ratchet wrench before you start cutting.

No-hub fittings should be checked for pinholes. Defective fittings can usually be returned to the supplies and exchanged for good ones.

CUTTING IRON PIPE   No-hub pipe comes from the factory in 10-ft. lengths, and is usually strangled and cracked to desired length with a soil-pipe cutter (snap cutter). I own two snap cutters, the Ridgid #206 and the Wheeler #527 (see pp. 88-89). The Ridgid #206 is the faster, better tool for cutting fairly long pieces of horizontal pipe in an unobstructed setting. The Wheeler #527 is better, though slower, for cutting vertical pipe and short pieces of pipe (as small as 3-in. pieces), and it can be used in much tighter confines than the Ridgid #206. The Ridgid #206 has the easier chain-wrapping design and is much quicker. The Wheeler has a cumbersome chain-wrapping mechanism but its chain is smaller, and will fit in tighter spots. With either tool, it's a good idea to make some practice cuts to learn how the tool works.

The Ridgid #206 has a ratcheting handle, so the first thing you do when using this model is to determine which direction you want the handle ratcheting in. Pushing downward on the handle is more comfortable; pushing upward is more strenuous, but in some circumstances more controlled.

Once you have determined the direction you want for handle movement, you have to wrap the chain around the pipe. When you do this, you should open the jaws all the way. You have to lift and either put the arrow on the ratchet pawl to the "open" setting or leave it halfway between "open" and "cut" in an unmarked neutral position that allows you to open and close the jaws without the pawl being engaged. Then you try to get the closest chain lugs into the up-turned "claws." The claws are spring loaded, so you use your fingers to press them up against the pipe, which closes the distance to the closest lugs. When you find yourself between lugs, you go to the next farthest ones. Once they nest in the claws, you tighten up the chain with the round adjusting knob until you can no longer turn it by hand, then set the arrow on the pawl to the "cut" mark. Start ratcheting the handle until the pipe pops in two.

The Ridgid #206 has one major drawback: If you do not open the jaws all the way each time before finding the closest chain lugs, you can put the wrong lugs in the claws and then bind up the ratchet and jaws before the pipe snaps, which renders the cutter useless. You have to send it back to a repair facility for dismantling to free up the jaws and ratchet. Years ago I painted white, exterior-enamel primer on the lugs that fit 2-in., 3-in. and 4-in. iron pipe. When I maintain these marks, I open the jaws just far enough to get the particular lug that I want into the claws. This simple measure saves a lot of time and prevents a costly repair bill.

The Wheeler #527 is a much simpler tool than the Ridgid #206. The Wheeler's scissored jaws are activated by turning a handwheel and screw, which pushes apart the jaw sections behind the axis point. In front of the axis point, the jaw ends host the chain-lug claws and one end of the chain. The claws on this tool curve inward (the opposite of the Ridgid #206). To get the claws over the chain lugs using this tool is more cumbersome and generally more time-consuming. Trial and error is the only way.

Once you have the chain wrapped around the pipe and in the claws, turn the handwheel clockwise as far as it will go. The Wheeler #527 uses a separate handle to tighten the chain. Apply either the long handle ratchet that comes with the tool, or use a large (16-in. to 18-in.) adjustable wrench.

Align the soil-pipe cutter on the mark (shown here is the Ridgid #206).

With the ratchet pawl on the "neutral" setting, tighten the chain of the soil-pipe cutter with the adjusting knob.

Today, iron pipe is generally cut with a soil-pipe cutter, or snap cutter, though it can also be severed with a grinder or a circular saw with a carborundum blade. The only tool needed for no-hub joinery is a tee-handled coupling wrench to tighten the clamps on the coupling. The same tools can be used with terra-cotta pipe.

SOIL-PIPE CUTTERS

I use two different designs and brands of soil-pipe cutters in my work. My front-line warrior is a Ridgid #206, which I would venture to guess is the most popular nationwide; I also own a Wheeler #527. The Ridgid #206 has a long handle that ratchets the cutting chain tighter and tighter until it cracks the pipe. This cutter works fine when you are trimming pieces 5 in. or longer from a horizontal full-length piece or a piece at least 3 ft. to 4 ft. long. However, it will not consistently cut 4-in. dia. iron pipe any shorter than about 4 in.

The Ridgid #206 is used to cut pipe 2 in. in dia. or larger. It is a very cumbersome tool to use on vertical pipe because its long, heavy handle will be banging you in the head while you are trying to wrap the chain around the pipe. If you are trying to get the chain through a narrow gap of structural members, you will have to hold the chain almost horizontal, which requires a good deal of strength. This is one ap-

For working with iron pipe or terra-cotta, you will need a soil-pipe cutter, no-hub or Calder couplings, and a tee-handled coupling wrench.

plication for which I greatly appreciate my Wheeler #527.

The Wheeler #527 has a short, stationary handle used solely for carrying the tool and holding on to it while you get the chain wrapped around the pipe. There is a round, scalloped wheel-knob with a squared male drive sticking out the top that you turn to open and close the jaws. The squared drive is the drive end of a long threaded shaft that passes through one scissored leg and nests in a boss on the opposite scissored leg. You open the jaws all the way and try to get the closest chain lugs into the claws, then tighten up the chain with the handwheel before using a ratchet to continue strangling the pipe until it breaks in two.

The Wheeler #527 is considerably lighter than the Ridgid #206, and the Wheeler chain is also lighter and smaller. Getting the lugs into the claws on the Wheeler is a lot more exasperat-

ing though, in part because the claws are rigid and do not swing. The separate ratchet that came with the Wheeler #527 when I bought it broke early on (it was made in Taiwan). I later fixed it, but not until I got used to using a 12-in. crescent wrench on the squared male drive.

The Wheeler #527 snaps the pipe with only the force of the constriction of the chain. This means that you can cut as little as 3 in. off the end of no-hub iron pipe. The Ridgid #206 applies an additional longitudinal shear, which can cause poor-quality cuts or prevent you altogether from successfully cutting very short pieces of pipe. The smaller-diameter chain on the Wheeler #527 can also many times fit around a wye or tee fitting, directly below the branch, which is corked (leaded) into a hub directly below or downstream. This enables you within minutes to attach a no-hub coupling; to do this work with hot lead would require hours.

With the Ridgid #206, which side of the pipe you place the chain connection on is an important decision, because it determines whether you will be shoving down or lifting up on the handle from the same standing position to crack the pipe. Shoving down on the handle tends to lift the pipe in front of your cutter and force the end of the pipe in back of you into the ground. If the pipe in back of you extends back a sufficient distance (3 ft. to 4 ft. or more), the pipe will stay flat on the ground and resist the downward force of the handle, keeping the energy of your movements in the chain. If the length of pipe is too short, when you try shoving down on the handle, there is not enough length to offer resistance, the pipe lifts up in front and no energy goes into the chain. Before I bought my Wheeler #527 cutter, I would slide a long digging iron through the no-hub pipe and then stand on the digging iron. This works, but you tend not to get very true cuts because of the tremendous shear force on the pipe too close to the point where you want to sever it.

GRINDERS AND CIRCULAR SAWS

A grinder with a large-diameter, flat cutting disc can be used for cutting iron pipe, but I don't cut pipe this way because of the risk of injury if the spinning disc

A grinder can be used to remove a small amount of iron pipe.

should fracture. Instead, I use a worm-drive circular saw with a metal cutoff blade for the occasional one-shot cut, if there is enough space to work. If you plan to do a lot of iron work, you might consider purchasing an abrasive cutoff saw for cutting pipe. You will still need the snap cutter for repair work, however.

COUPLING WRENCH

Both no-hub iron and terra-cotta are joined with rubber couplings that have clamps around them that must be tightened. The tee-handle wrench has a preset clutch that releases when you have tightened the nuts on the coupling's hose clamps. There are two popular designs on the market: one is the Pasco #7020 and the other is the Ridgid #902. Both wrenches ratchet only in the tightening direction. The big difference between these designs is in the method in which they

loosen the hose clamp. The Pasco wrench has a little fold-out wing on the nut shaft that provides leverage to back up the hose-clamp nut. This is a cumbersome maneuver when the no-hub coupling is down in a recess and you cannot easily reach it with both hands. The Ridgid wrench has a locking feature that lets you back up the nuts using only one hand. The Ridgid wrench costs two or three times more than the Pasco wrench. It is also top-heavy and prone to being dropped (inside walls where you can't get it back). The Pasco wrench is smaller and more comfortable to operate in the forward mode.

Some codes specify how much torque should be applied to the couplings. The Pasco #7020 has a preset clutch that releases at an acceptable torque.

When you get near the point of popping the pipe (you have to guess when this point is), you can either grab the tool's rubber-clad handle to keep it from dropping or let it fall with the pipe if that seems safer. For the first few practice cuts, you should hang onto the tool handle.

Very rarely does old or new no-hub iron pipe snap flush. Most of the time you will find little spikes, maybe ⅛ in. high, or V-shaped notches around the edge after snapping (a spike on one piece means a corresponding notch on the other piece). Occasionally you will find a larger (say ⅜ in.) spike or notch on the edge. In the case of a spike, you get out your 12-in. crescent wrench and slide the jaws down it to the base of the spike. Then tighten the jaws as tightly as you can. A firm, quick inward push on the handle will usually break off the spike close to flush with the remaining edge. Always try to remove the spike first, before cutting another piece. If the notch is no deeper than ¼ in., I use this end on my next measured piece. But if it goes deeper, I move back 3 in. or 4 in. and snap again, trying for a better edge. It is impossible to snap ⅛ in. or ¼ in. of pipe; removing that tiny amount calls for tedious work with a grinder.

In preparing to splice a fitting into an existing run of iron pipe, I mark the length of the new fitting on the existing pipe, using yellow keel (lumber crayon) or white grease pencil, and then I add ⅜ in. to allow for the fact that pipe rarely cracks flush. It also allows enough room for the ridge stop in the rubber collar to drop into. If the splice opening is too tight and you can barely fit the fitting in by itself, then there is not enough play for the ridge stop to hang down properly. It will cause a lump or bulge and the collar will not lie flat, which in turn means that the outer band will not apply force evenly on the mating parts. The likely result is a leak or a pipe/fitting separation later on.

Snapping a piece of 4-in. iron too short is something like cutting a header too short, so it's a good idea to follow the old carpentry rule: Measure twice, cut once. Seeing your mark through the chain of the snap cutter is not always easy. So you will find yourself sliding the chain back and forth several times before you feel that you have put the cutting wheels right on the money.

Snap cutters weigh a lot, and one of the trials of using no-hub pipe is having to lug this awkward tool around. Doing a few, quick cuts on a repair job is not so bad, but two or three long days with it can get tiresome, and making vertical cuts with the snap cutter is an outright form of masochism. But I'd take that any day over trying to cut the pipe with a reciprocating saw. Using a reciprocating saw with a carborundum-coated blade, however, you might spend well over an hour and use maybe three or four blades, at $15 to $20 apiece.

NO-HUB COUPLINGS   The no-hub coupling (see the sidebar on p. 20) is almost universally accepted for joining no-hub pipe and fittings inside the structure, above grade, and some communities may let you use it below grade. No-hub couplings can be purchased with different-sized diameters on each half of the coupling to join no-hub iron pipe and fittings of differing diameters.

The worm-drive hose clamps on each end of the metal band are held in place by one or more rivets or by a lapped, punched hole and no rivets. These clamps, if of good quality, may be unscrewed all the way and the end pulled free of the drive screw, and then re-inserted and once again screwed together. (Since you can easily cut yourself on the no-hub band, it's a good idea to wear leather gloves when working with these fittings.) Removing the clamps enables you to split open the band to unwrap it from the rubber collar. Because of this ability to open up, the coupling can function like a union, which allows you to work from different directions — a handy feature when you have to splice in a fitting on an existing pipe run. However, if you do not tighten the worm-drive hose-clamp screws alternately an equal number of times as you go, or if the pipe and/or fittings bridged by the coupling are of different diameters, the under-lapped section of the no-hub band has a tendency to squish out to one side and not make a good seal, in spite of the mated grooves.

There are distinct quality differences among the various brands of no-hub couplings with respect to the outer band and especially the worm-drive hose clamp. Good-quality couplings are all stainless steel, including the worm-drive screw, and have thicker-walled stainless-steel outside wrapping. Couplings of lesser quality use thinner wrapping, and the non-stainless-steel worm-drive screws might even be zinc plated. Good-quality couplings like those manufactured by Ideal and Norton have hose clamps that separate and re-engage smoothly. Lesser-quality ones cross-thread and bind up when you try to re-engage the end of the clamp into the drive-screw housing. Good-quality couplings have clamps that do not strip when you tighten them to the extreme; inferior couplings will "jump a groove" before reaching design torque and then never fully tighten again, rendering them useless. Good-quality couplings have a hefty, well-set rivet holding the clamp to the band. Lesser-quality ones have a tiny, poorly set rivet or lapped hole that breaks, allowing the hose clamp to fall off.

Before putting in a large order of couplings along with your pipe, buy just a few of whatever brand your supplier is selling, and then try to take them apart and put them back together again. Then place them on a piece of pipe or fitting and tighten them up all the way with the no-hub wrench until the clutch slips. If the clamp strips before you hear the clutch click or if you have other re-assembly problems, find another brand and test that one too.

The worm-drive screw head on most no-hub couplings is $5/16$-in. standard hex. Don't buy couplings with drive screws that are not $5/16$ in. The head may or may not have a slot for screwdriver application. This screwdriver slot can come in awfully handy when you have to undo a coupling that is buried in structural members or turned in a trench or a wall so as to deny access to the standard no-hub ratchet wrench. You can use an offset screwdriver to back up the screw a few turns until you can use your fingers to finish unscrewing the clamp. This situation often arises on remodel and repair work.

On your water test of the no-hub DWV piping system, you will find that when leaks appear they are most likely to be caused by pinholes in fittings, a cracked fitting or imperfect or out-of-round fittings. Many times you can stem the leak with a few additional turns of the screw with the 4-in. adjustable wrench, if you have so positioned the coupling as to provide easy access. So, as you assemble your no-hub DWV system, be constantly thinking about access to each coupling and position it accordingly. Assume that you will have a leak at each coupling and need to tighten it further or replace the fitting and/or coupling. It is rare not to have any leaks at first testing in an entire no-hub system.

There is an expensive attachment for an electric drill that drives both no-hub coupling hose-clamp screws simultaneously, but it does not fit well with many different brands of electric drills. I sometimes use a $5/16$-in. nut-driver bit in a 3-in. magnetic extension in one of my cordless drills to assemble the fittings to the pipe initially, then I finish tightening the couplings with the hand ratchet wrench. Unlike cementing ABS or soldering DWV copper, you rarely ever position the no-hub fitting just right the first time — there always seems to be a last-minute adjustment. That's where the cordless drill is so handy. It saves enormous amounts of time tightening and loosening the hose clamps.

The no-hub rubber collar has a few idiosyncrasies of its own. When trying to slip the coupling onto the pipe or fitting, you will in most cases have to work the rubber collar on first, independently of the band. There is a ridge running around the inside circumference of the collar. This ridge is a stop that ensures an even distribution of the coupling on the two mating parts. Always make sure that you have the pipe or fitting bottomed out against this ridge during assembly.

When working from one direction only, the rubber collar can usually be spread open sufficiently with your fingers to fit onto the pipe and fitting. If the couplings have been lying on their sides for a while, they may have sagged into an oval shape, and it might take a bit more effort to get the rubber collar and outer band in place.

When attempting to splice a no-hub coupling into an existing pipe run, you will find that the little inside rubber ridge stop in the collar prevents the collar from sliding easily onto the pipe or fitting. You can get the coupling on using a pair of pliers to grip the collar and then pulling firmly at different points around its circumference. An easier method is to position the rubber collar onto each end of the fitting, up against the ridge stop. Then with the fitting pressed between your legs, use both hands to fold the outside half of the rubber collar back on itself. Keep shoving until the ridge itself goes on farther, past the edge of the fitting. Once folded back, it will stay there.

Next slide the stainless-steel bands onto the pipe ends, leaving room for the rubber to be rolled back onto the pipe. If you have another immediate fitting, you might have to leave the outer band off for now. You'll have to tug the rubber collar toward the pipe until the ridge stop falls into the gap between the pipe and fitting and the collar lies flat, all the way around. Now you slide the outer band over the collar and line up both edges with each other.

If the loose band doesn't form a good circle, you cannot easily slide the band over the collar. Make sure that you have backed up the screws until the hose clamp is open all the way. By gripping the edges of the band, you can usually pull it open enough to slide it over. This can be somewhat exasperating, depending on what position your body is in and where the section you are working on lies. If the band will not slide over the collar because it is misshapen, back off on the drive screws until the ends of the hose clamps pull free from the screw. Now you can peel open the band and easily place it over the collar. The end of the band with the mitered corners goes underneath the square edge. This helps the band resist buckling when you tighten it up. Now restart the hose clamps back into the worm-screw housing and snug up the band.

When joining no-hub iron pipe and fittings to other pipe and fittings of different materials, you can use adapters made for those materials that ensure that both materials are the same diameter at the joint and that the no-hub coupling is equally compressed on each end. These are called no-hub adapters regardless of what material they are made from. In some locales, no-hub couplings on iron pipe are limited to use inside the building; they can not be used in the ground, as in the case of slab.

MISSION COUPLINGS   Because of its thicker outer stainless-steel wrap, the Mission coupling is stronger than the no-hub coupling, and some local authorities require the use of the Mission coupling for burial of no-hub pipe and fittings below grade. However, the Mission coupling may not be as easy to install. You can undo the clamps and split open the Mission coupling, but depending upon the nut-retention design of the lugs, you may lose the nuts if the design of your coupling allows them to come loose when the clamps are separated. Also, when this coupling is equipped with draw-bolt type clamps, it often cannot be tightened enough to stem pesky leaks on water test because the lugs on the clamps act as a stop when they come to rest against the bolt heads when you are going for that last bit of torque. Mission couplings with worm-drive hose clamps work just as well and as easily as no-hub couplings.

Mission specializes in couplings for joining pipe and fittings of different materials and different outside diameters (4-in. iron by 3-in. plastic, or 3-in. plastic by 2-in. copper, 2-in. iron by 2-in. plastic, 2-in. plastic by 1½-in. DWV copper, etc.). On these fittings, the rubber collar is a different inside diameter on each side of the ridge stop. Mission couplings are tightened with a tee-handled coupling wrench (see the photo on the facing page). These couplings are used primarily in remodeling, which is discussed in Chapter 9.

ALL-RUBBER COUPLINGS AND FITTINGS   Fernco and Indiana Seal manufacture all-rubber couplings with stainless-steel worm-drive hose clamps, but without an outer stainless-steel wrap. That makes for a big difference in strength and what applications you can use them for. These and standard no-hub couplings are not the same. The all-rubber couplings are made in increasing sizes for joining different diameter pipe and fittings within the same material, and for differing diameter pipe and fittings of different materials. For increasing sizes, the coupling is often molded to the two different diameters without the need for separate bushings.

This type of all-rubber coupling is a wonderful invention, but not all locales will let you use them. Those communities that do sanction them usually allow them only for building sewers, outside, in the ground, even though they might physically fit in the wall on vent and drain piping. Because of their one-thickness, all-rubber construction, these couplings cannot resist a lot of shear force on the joint, as you could experience with horizontal iron piping hung from structural members. If you use the rubber-only coupling, make sure that you brace the two pipes that you want to join so no shear forces are applied to this

The clamps on a Mission coupling are tightened with a tee-handled coupling wrench.

joint, or the rubber may eventually tear or come loose and the joint will probably leak. I suspect that this is why they are mostly allowed only for burial, where the piping is supposedly totally supported along its bottom length and the confines of the trench limit its sideways movement.

Some companies, including Mission, make entire fittings out of rubber — sanitary tees, combination wyes, 90° elbows, 45° elbows and even trap J-bends. These are frequently not sanctioned by local authorities, regardless of what major codes say. But they can be very useful on buildings whose remaining life can be counted on two hands, and I wouldn't hesitate to use them in this situation. I have also used them during the demolition stage of remodel projects when temporary hook-ups are needed before the new work begins. The tubular-brass sized rubber J-bend works well for temporarily hooking up a bathtub for unfortunate folks who are "living in the dust" while their home is in transition. I once used one of these traps for about a week on a tub that was carried out each day and put back in the evening.

CIT COUPLINGS   In locales that do not sanction the joining of no-hub iron pipe and fittings below grade outside of the structure with either standard no-hub couplings or the Fernco and Indiana Seal all-rubber couplings, you must use all-rubber CIT (cast iron to cast iron) couplings. This all-rubber coupling for 2-in. to 4-in. cast iron pipe is made from a thicker, tougher rubber material than the Fernco and Indiana Seal couplings. Its raised center cross section gives it added strength and allows for slight pipe movements or misalignments in the ground. The plumber can even make very slight changes of direction in a long pipe run by "tweaking" each length of pipe just a bit. This is especially helpful when the pipe run is losing elevation fast. It can save you from using two iron fittings to compensate for the drop.

JOINING TERRA-COTTA PIPE

Terra-cotta is seldom used these days, but if you work with it you'll need a soil-pipe cutter, a reciprocating saw with a Lenox 8-in. carborundum-coated blade, a white grease pencil or soapstone pencil, a no-hub coupling wrench, a mini-hacksaw, a razor-sharp pocketknife and a whetstone to keep it sharp. Wear gloves when you handle terra-cotta, and use protective eyewear when you tap on it or break it up. When new terra-cotta pipe is snapped with the soil-pipe cutter, the edges can be razor sharp. There is no such thing as a superficial cut from this stuff. It always bites with a vengeance.

The new terra-cotta pipe used in residential work comes from the factory with a Calder coupling attached to one end. The Calder coupling is a rubber sleeve, similar to the sleeve on a CIT coupling, with a worm-drive stainless-steel hose clamp on the open side. A nice feature of this coupling is that it allows the pipe to move about somewhat in the ground, relieving stresses that might otherwise break its ends. It also allows you to make slight, intentional changes in direction. Over a distance of several pipe lengths, you can make your pipe line swerve to the left or right or slope downward without using any fittings.

The walls of the Calder coupling are too thick to roll back on itself, but as long as you are working in one direction, you will have little difficulty nosing the pipe and fittings into successive couplings. However, the work can be more difficult if you have to splice a pipe section or fitting into an existing pipe run, as described on pp. 94-95.

Like the pipe itself, terra-cotta fittings also come with factory-attached rubber couplings. (A wye or a tee would have two pre-installed couplings.) Because these fittings are very bulky, often you have to trim off some of their length to get them fitted in tight spots. Aside from the fragility of the material, this is the next biggest argument against using terra-cotta.

CUTTING TERRA-COTTA PIPE

Terra-cotta pipe is a dream to snap. I use the Ridgid #206 snap cutter (see pp. 88-89) for 6-in. and 8-in. dia. new pipe, which I usually cut on the ground alongside the trench. Because I rarely cut terra-cotta pipe, I don't mark the lugs ahead of time; I just keep a close eye on the gap between the ratchet end of the

New terra-cotta pipe can be cut with the Ridgid #206 soil-pipe cutter.

handle and the top jamming end of the jaws. If I run out of room before a snap occurs, I know that I chose the wrong lugs.

I use the Wheeler #527 to cut old excavated pipe because I do not want forces applied anywhere other than to the chain itself. Terra-cotta pipe is extremely sensitive to off-center forces, and you can loosen up joints that are still buried back 3 ft. or 4 ft. from where you are working if you use the Ridgid cutter on pipe that has no soil support for 2 ft. or more. Snapping terra-cotta takes only a fraction of the number of arm movements on the snap cutter's handle or wrench that snapping no-hub iron requires. Mark your cuts with yellow keel, grease pencil or soapstone. On new terra-cotta pipe, which is reddish orange, yellow keel won't show up well; it's better to mark with white grease pencil or white soapstone.

As terra-cotta ages, its color changes (see the sidebar on p. 25), and so do its working properties. Old, in-service terra-cotta pipe that has a greenish-yellow hue generally doesn't snap well because the bottom third of the pipe, which can become saturated with water, handles differently from the top two-thirds, which is usually dry. If you try to sever pipe in this condition with the soil-pipe cutter, you will crush or cave in the bottom portion and crack the pipe for several feet back. The top section of the pipe will not part, at least not where you want it to. If you have dug down several feet or more to reach the pipe and this fate befalls you, you might have to dig another 3 ft. to 5 ft. farther back to reach a portion of pipe without cracks. That's a lot of earth to move. So I now play it safe with greenish-yellow pipe and use my reciprocating saw with the longest carborundum-coated blade I can find (Lenox #0577-800 RG). I'd rather spend 45 minutes sawing for a sure thing than risk messing up while snapping and then have to spend hours and hours more on the job.

When snapping chestnut-brown terra-cotta that has been in the ground for years, go slow because you may not hear a definite crack, thud or pop. The pipe may snap almost silently, and if you keep going with the snap cutter you might crush the pipe that you will be wanting to fasten to. Here's a tip for removing a damaged section that will minimize the risk of damage to the surrounding pipe. After marking the pipe for the distance between cuts on a splice job, first make a cut in the middle and then cut on the marks. If the cut sections do not fall out or gently tap out, then cut one of them in half. Now you'll have a third piece that you can tap on more aggressively without as much danger to the sections that are to remain intact.

MAKING REPAIRS

To splice in a piece of pipe and/or fitting into an existing terra-cotta pipe run, you need to remove the coupling, then take the clamps off it and turn the coupling inside out. This puts the ridge stop on the top. Then, with a very sharp pocketknife or a utility knife with a new blade, you should cut off the ridge stop (see the photo above right). Over the years I have developed a technique for doing this. With one leather-gloved hand (you should be sitting down) squeeze the coupling (with hose clamps off), which raises the center, where the ridge stop is. Now with the knife, start cutting into the rubber where the ridge meets the wall of the coupling. Go in with the tip of the blade at a slight downward angle. Don't try to cut real deep. Go around twice lightly, just to make a track, then turn the coupling around in your hand and do the same to the other side of the ridge stop. The stop parts from the coupling as one complete rubber ring. You'll have to keep a squeeze on the arc that you cut with the knife. If you don't, the blade

The ridge is cut off the inside of a Calder coupling so it can be used as a repair fitting in terra-cotta pipe runs.

will be pinched by the rubber and it will not pull very easily, and you will not have as much control of the blade.

Once the stop is cut out, turn the coupling right side out and replace the hose clamps. On new couplings the hose clamps often come with the drive screws facing opposite directions. When you put them back into the grooves, put them on so that both screws face the same direction. Often you won't be able to get at both screws equally well or at all, if you leave them as you originally found them. For the spliced piece of pipe or fitting, you will need two of these customized Calders. After cutting out the section of pipe run to be replaced, slide one Calder onto and down the excavated pipe ends, leaving the cut edge flush or a little past the coupling. Now you can insert the new piece of pipe or fitting and then slide the couplings back, positioning their width equally onto the existing pipe and the new splice.

If you dig up a terra-cotta line and are going to replace all or part of it with new terra-cotta, then break up the old line into small chunks and sprinkle a lot of it into the backfill. It's an old plumber's practice for warning future excavators that they are approaching a fragile pipe run. Terra-cotta is destroyed by pickaxes and even shovels. Usually an excavator who encounters broken pieces of old terra-cotta will slow down and proceed with caution before scoring a direct hit on the replaced section.

DRAIN, WASTE AND VENT LINES

With the preliminaries of selecting the pipe and fittings and sizing the plumbing runs out of the way, the layout and installation of the pipes can begin. Because the DWV system relies on gravity to work effectively and because its pipes are larger than water-supply pipes, it requires the most direct routes from fixtures down to the building drain and up through the roof. Therefore it is installed first. In this chapter we will consider three critical aspects of the DWV system design — the building drain, fixture vent-to-drain patterns, and the vents — then discuss the installation of the system with the two types of construction: off the ground and slab.

Begin your assessment of a proposed DWV system with a visit to the site, which will tell you a lot more about possible problems than the house's blueprints alone. If you will be plumbing an entire house (or an addition that will occupy new space on grade), it sometimes helps to see the actual size of trees that show up on the plot plan. Is there any poison oak or poison ivy? Are there any retaining walls, tree stumps or large rocks that would spoil the path for a good, true-running building sewer to the city sewer inlet or septic inlet? It is also important for you to know how accessible the site is to your vehicle. You'll be bringing a lot of materials to the job, and having to lug pipe and fittings any distance on foot is a lot more difficult and more time-consuming than just dumping them off the truck.

If there is a city sewer inlet to attach to, you usually can find out its depth by checking with the local public-works department, but I like to excavate it anyway and see where it is with my own eyes. Once the inlet is excavated, you can measure its depth below grade, the height of the finish floor above grade and the distance from the sewer inlet back to the foundation perimeter. These important measurements tell you how much vertical working space, or "altitude" (see the discussion on pp. 97-98), you have to install the below-grade drainage lines and the building sewer.

When you have sufficient altitude on ground-floor drainage lines and the building sewer line, you can create a main drainage system that operates on gravity: The weight of water alone will carry liquid and solid wastes out of the house's drainage system, through the building sewer and into the city sewer or septic system. If all that had to be done without the aid of gravity, it would be very costly; other than the cost of the water, a gravity DWV system is free. A

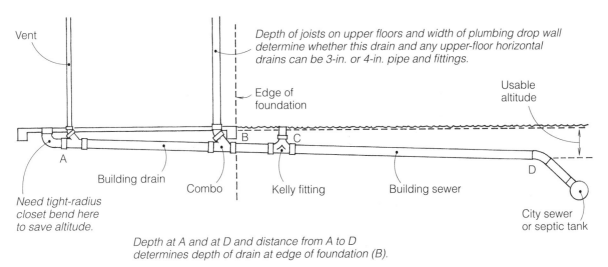

Vent

Depth of joists on upper floors and width of plumbing drop wall determine whether this drain and any upper-floor horizontal drains can be 3-in. or 4-in. pipe and fittings.

Edge of foundation

Usable altitude

B

A

C

Building drain

Combo

Kelly fitting

Building sewer

D

Need tight-radius closet bend here to save altitude.

City sewer or septic tank

Depth at A and at D and distance from A to D determines depth of drain at edge of foundation (B).

The depth at C and at D and the distance from C to D tell you how much slope the building sewer can have.

gravity DWV system requires a minimum downward slope of ⅛ in. per foot, which all codes demand. At lesser slopes, the velocity of liquid wastes flowing through the pipes is too slow and too weak to carry the solid wastes along. At slopes greater than ½ in. per foot, the liquid waste moves too fast for the solid wastes, and the latter can be left behind. The optimum downward slope of drainage pipes is ¼ in. per foot. At this slope, the solid and liquid wastes stay together and flows are at their best. When the drainage piping needs to lose altitude more rapidly, it is best to keep the pipe running at ¼ in. per foot and then send it downward for a short distance at 45°. At this pitch or greater, the solids slide downward without any difficulty.

If you are plumbing a house or an addition whose ground floor is below this ⅛-in.-per-foot downward-slope threshold, you'll need to install a pumped drainage system. The drainage piping for the structure is installed just as it for a gravity system, but the effluent flows into a sealed collection basin that contains a sewage ejector (a pump equipped with a liquid level sensor that turns the pump on when the basin becomes full). The sewage ejector pumps the liquid and solid waste under pressure from the col-

lection basin to a convenient location and altitude, where it can once again can resume its journey to the city sewer or septic system by gravity. This device requires an electrical circuit, and depending upon the size of your home and the capacity of your pump, providing the electrical circuit can cost as much as or more than the ejector itself. For more on sewage ejectors, see pp. 197-200.

ALTITUDE

Altitude is a critical concept in plumbing. Altitude is the available depth you have to work within, for any given distance, when designing vent-line to drainage-line connections with pipe fittings and establishing acceptable slope on drain and sewer lines. In order to lay out pipes so that the DWV system functions efficiently, you will need to know the available altitude for the distance between the city sewer or septic and the edge of the building and for the span between the uppermost toilet and the edge of the first-floor foundation. You will also need to know the available altitude in the depth of upper-floor floor joists over the

Where altitude is limited, it may be better to use one piping material over another. For example, a wye to 45° bend joint in ABS plastic (left) takes up less altitude than the same joint in no-hub iron (right).

distance in which drain lines have to run to a plumbing wall, where they will drop down. It is within these altitudes that you must execute your piping designs.

Limited altitude in a particular location may restrict your choice of fittings and materials. For example, the photo above shows a wye with a street 45° bend in ABS (at left) and a wye with a 45° bend in no-hub iron (at right). Because the plastic fittings nest and the iron fittings butt, there is a difference in the altitude of the joint.

With DWV plumbing, the more altitude, the easier your work will be. The plumber's biggest enemy is the absence of altitude.

THE BUILDING DRAIN

The building drain is the main drain in the house, and all other drains and waste lines flow into it. Its configuration depends on the location of the fixtures in the house. It may be one continuous run of pipe from a far-flung toilet to the edge of the foundation where the sewer awaits, with the other drains joining it in short, direct runs. Or it may go straight down the center of the house, with other drains making long, below-floor, horizontal runs back to it from ex-

terior walls. Sometimes you end up with three large-diameter drains that start out as toilet drains at their upper ends and converge within a few feet of a very short building drain. You could plumb dozens and dozens of custom homes and never have the same configuration. Two possibilities are shown in the drawing on the facing page.

When designing the building drain, give priority to toilets, which discharge most of the solid wastes. (The smaller drains for showers, tubs and sinks are less important to consider at this stage.) Begin by visualizing the most direct link for toilet vent-to-drain patterns (see pp. 101-105) and mentally continue to draw the building drain traveling to the building-sewer connection at the edge of the foundation. Use a sketch pad and pencil to record your thoughts. You should be able to weave the small-diameter branch lines to the building drain with fewer than 135° of aggregate directional change. If you need to assign secondary priorities to the branch lines, give priority to tubs and showers because you cannot place an upper-terminus cleanout for these fixtures easily without a floor-accessed cleanout with a floor plate over it (a solution that is too "industrial looking" for residential construction). If one vent-to-drain pattern turns out to be a problem, try to modify it.

BUILDING-DRAIN CONFIGURATIONS

TOP VIEW, VENTS NOT SHOWN

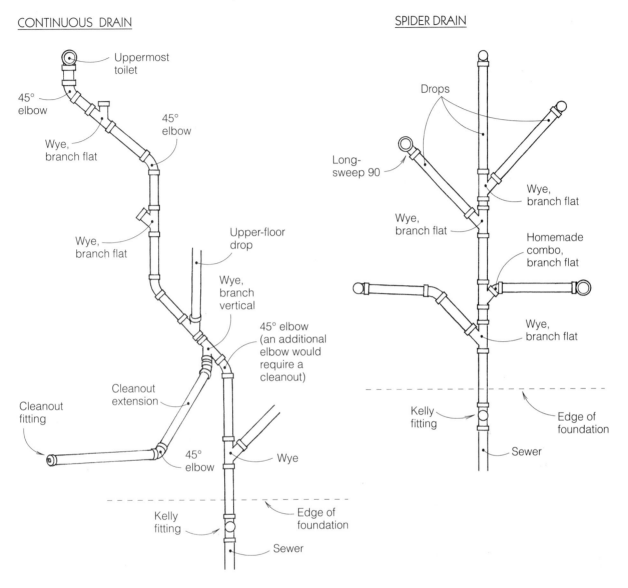

CONTINUOUS DRAIN

- Uppermost toilet
- 45° elbow
- Wye, branch flat
- 45° elbow
- Wye, branch flat
- Upper-floor drop
- Wye, branch vertical
- 45° elbow (an additional elbow would require a cleanout)
- Cleanout extension
- Cleanout fitting
- 45° elbow
- Wye
- Kelly fitting
- Edge of foundation
- Sewer

SPIDER DRAIN

- Drops
- Long-sweep 90
- Wye, branch flat
- Wye, branch flat
- Homemade combo, branch flat
- Wye, branch flat
- Kelly fitting
- Edge of foundation
- Sewer

When you get a scheme that works, then flesh out the connections. The major plumbing codes require 45° wye fittings to join horizontal to horizontal and vertical to horizontal drainage piping. So visualize wyes with the branch inlets swept the proper direction to compensate for flow as you prospect for possible connection points along the imaginary path of the building drain as it exists in your sketch. Horizontal branch lines approaching the branch inlets of these wyes must be sloped, preferably at ¼ in. per foot. That means lifting the branch inlet of the wye slightly so that it is above flat and level; higher, up to 45°, is better. Dropping a branch line into the vertical inlet of a combo wye is the best way to go. Changes in direction in small branch lines are best accomplished with 45° elbows and 22½° bends.

LOCATING CLEANOUTS (TOP VIEW)

TOP VIEW

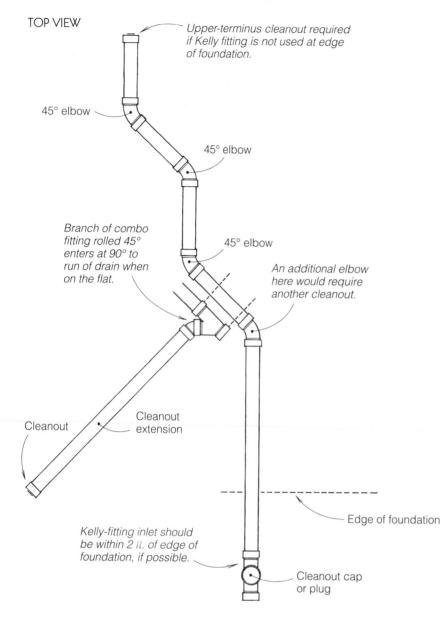

Upper-terminus cleanout required if Kelly fitting is not used at edge of foundation.

45° elbow

45° elbow

45° elbow

Branch of combo fitting rolled 45° enters at 90° to run of drain when on the flat.

An additional elbow here would require another cleanout.

Cleanout

Cleanout extension

Kelly-fitting inlet should be within 2 ft. of edge of foundation, if possible.

Edge of foundation

Cleanout cap or plug

Try to avoid designing a horizontal drainage run that requires a 90° turn. If site and structural circumstances make one necessary, you'll usually be required to use a combination wye and ⅛ bend (combo) with a cleanout hub and plug in the upstream barrel. Local codes differ on this, but you might also have to install a cleanout extension line to the out-side edge of the building. Most local authorities will accept a ⅛-in. or ¼-in. per foot slope downward on a cleanout extension line. If a Kelly fitting (see p. 21) is used on the building drain just outside the structure, no upper-terminus cleanouts are required. Cleanout requirements are summarized in the drawing above.

FIXTURE VENT-TO-DRAIN PATTERNS

One of the most important aspects of designing a drainage system is choosing the fittings that will allow each fixture's vent to connect to its drain within the available altitude. I call these connections vent-to-drain patterns. The various fixture vent-to-drain patterns discussed below are code-approved configurations of fittings and pipes between the fixture and the vent/drain junction. When I am studying the possibilities of various drainage designs, I visualize these patterns to see which ones will work best with the fixture plan for the house. Some vent-to-drain patterns require more altitude than others, some require more side room.

For sinks of any kind, there is only one branch fitting: the sanitary tee. The vent attaches to the top inlet; the trap arm attaches to the branch inlet. For tubs, showers and toilets, however, there are choices. The tub and shower will use one of two branch fittings, usually a wye, but sometimes a tee (with your local authority's blessing). The tee can be a sanitary tee or a vent tee. The toilet may use the same two branch fittings, or possibly an additional fitting, used only with it: the low-heel vent 90°. Tubs, showers and toilets can use the sanitary-tee branch fitting vertically, like a sink, though I find more applications for the branch fitting with the run horizontal.

For the vent line to be effective on a horizontal drain line, it must connect to the drain line above the flood level of the drain. At most, any horizontal drain line is about one-third full when conveying waste (you don't worry about vertical drainage lines). This means that when you position the inlet branch vertical, the waste runs well below the open inlet. Since you have so much altitude here, you can roll the branch to one side or the other of top dead center by as much as 45° and still be above the level of the waste running below. To stay on the safe side, most codes state that you cannot exceed this 45° roll.

If you have very generous altitude above the fitting, you can leave the branch inlet perfectly vertical and come up with a vent line as far as you can before encountering interference from structural limitations; at that point you would need to use a 90° elbow to change direction to the horizontal (¼ in. per foot upward slope), then take the vent to a convenient wall to turn up and head for the roof. When you have less altitude, you roll the branch inlet to the side that will send the horizontal vent run in the direction you need to reach the turn up. With a vertical branch inlet (sanitary tee) and a 90° elbow on top, you can swivel the vent run in almost a full, flat circle. If you roll the branch inlet, the 90° elbow will no longer work at any angle, but only in a parallel run, upstream or downstream with the drain line. If you roll the branch inlet to 45° either side of top dead center and do not use the 90° elbow to parallel the vent to the drain line, you then are limited to sending the vent run off at 90° to the drain line by using a 45° elbow. You can cheat somewhat and add a 22½° bend to the 45° elbow when it is swiveled slightly to head upstream or downstream.

When you use a sanitary tee, you use the fitting's top inlet as the upstream inlet because of the branch's sweep. You need the entering air to be traveling the direction of the waste flowing down the drain line. For the same reason, when you use a wye, the branch inlet must be pointed upstream of the drain so that the air coming down the vent is following the direction of the waste flowing in the drain.

Because the toilet drain is the largest-diameter drain pipe in your DWV drainage system, it should be designed first. Then you can concern yourself with the smaller-diameter drain lines of lavatories, kitchen sinks and tub/showers.

FLOOR-MOUNTED TOILET PATTERNS

Begin the design process by visualizing a bathroom with a conventional, floor-mounted toilet positioned near one wall. The back of the toilet's tank is an inch or less from the wall that will contain the supply piping (I call this the plumbing wall or wet wall). Also visualize the vent wall, the wall in which the toilet's vent travels upward toward the roof or another properly sized vent. This may or may not be the same wall as the plumbing wall.

Depending on circumstances, there are six possible vent-to-drain patterns for the toilet. For simplicity's sake, I will describe their makeup using ABS or PVC plastic pipe. The patterns can be closely replicated, dimension-wise, using soldered DWV copper, which does not have any vent fittings. They can also be closely replicated in no-hub iron, because in larger-size fittings, no-hub is close to the dimensions of plastic fittings. However, no-hub has no vent fittings, and where you can use a physically smaller street or vent fitting in plastic or street fitting in copper, the no-hub version will lose available altitude because its standard fittings are larger.

FIRST PATTERN   The first pattern is used when a toilet drain line approaches the plumbing wall at 90° and there is less than generous altitude (2x10 joists). Connections are a 4-in. 90° elbow (street, ¼ bend, mid or long sweep, depending upon how much altitude you have) with the closet flange at its vertical, upper inlet; a 4x2 wye, and a 2-in. street or hub-to-hub 45° elbow (⅛ bend). There is also a piece of 2-in. pipe to get over to and under the vent wall, which in this case is also the plumbing wall, and a vertically upturned 2-in. 90° elbow (of any radius) to carry the vent upward. The wye fitting is rolled 45° to save vertical altitude, with the 2-in. inlet at 45° to either side of center, the maximum degree allowed. The wye branch's sweep also points in the direction of the

drain line's flow, even though only air is flowing through the vent pipe. If altitude is a concern, this pattern can also be done in 3-in. pipe. You would still use a 45° elbow on a 3x2 wye, though.

SECOND PATTERN   The second pattern, which has a sanitary tee instead of a wye fitting, is a variation of the first pattern and can be used in similar situations, if your local inspector has no objection. The substitution of a sanitary tee for a wye has no bearing on drainage flow, but some inspectors may favor the wye. As in the first pattern, the tee branch is rolled 45° to save altitude. A vent, street or ¼ bend 90° elbow runs the vent to the wall, and the upturned 90° elbow continues the vent upward. This pattern can also be run in 3-in. pipe, with a 3x2 tee.

PATTERN 2

Drain line approaches plumbing wall at 90°; vent joins drain at sanitary tee.

PATTERN 1

Drain line approaches plumbing wall at 90°; vent joins drain at wye fitting.

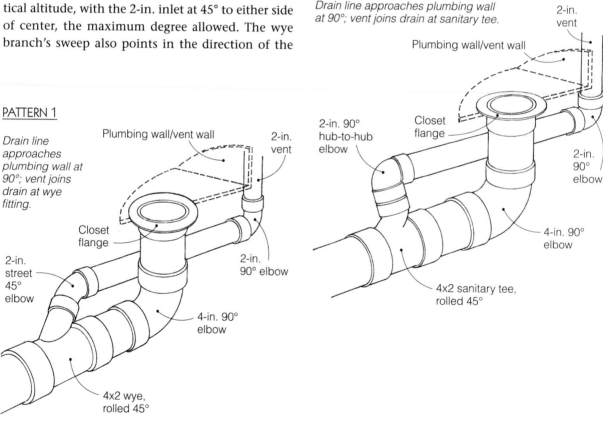

THIRD PATTERN The third pattern is used when a toilet drain line parallels the plumbing wall, which is also the vent wall. In this pattern, a 4-in. 90° elbow joins the top inlet of a 4x2 wye. The wye branch is rolled to 45°, and an additional 45° elbow and a 22½° bend (⅟₁₆ bend) coupled to it straighten out the vent line going over to the plumbing/vent wall. I would use this pattern to nudge the vent a few extra inches to get it into a stud bay if I couldn't position the wye farther downstream on the drain line due to structural limitations.

PATTERN 3

Drain line parallels plumbing wall; vent joins drain at wye fitting.

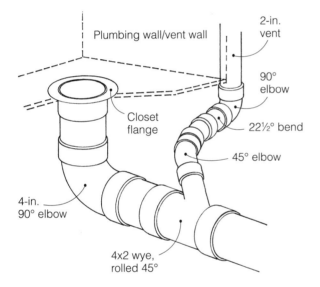

Plumbing wall/vent wall

2-in. vent

Closet flange

90° elbow

22½° bend

45° elbow

4-in. 90° elbow

4x2 wye, rolled 45°

FOURTH PATTERN Sometimes the framing of the plumbing wall behind the toilet is not well suited for a vent run, perhaps because it holds heating or air-conditioning ducts or electrical wiring. In this situation, it's best to choose another wall for the upward vent run. The fourth pattern will work with a drain line approaching the plumbing wall at any angle. This pattern uses a 3x2 or 4x2 sanitary tee with the branch vertical, to which is attached a 90° elbow. Using a 90° elbow of any radius will work in generous altitude situations; a street 90° elbow saves as much altitude as possible. The elbow fitting lets you swing the hub (and the horizontal vent line that connects to it) almost 360° in the direction of whatever vent wall you have chosen. This vent line should slope upward at ¼ in. per foot. An upturned 90° elbow of any radius is installed under the chosen vent wall, and the vent is started up inside the wall toward the roof or other properly sized vent. I use this pattern when the drainage line is neither parallel nor perpendicular to the plumbing wall.

PATTERN 4

Drain line approaches plumbing wall at any angle; vent joins drain at any angle..

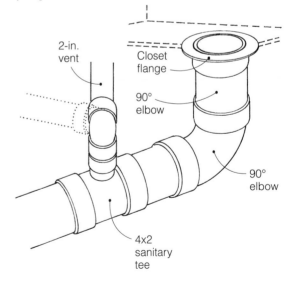

2-in. vent

Closet flange

90° elbow

90° elbow

4x2 sanitary tee

FIFTH PATTERN   The fifth pattern employs the low-heel vent 90°, which comes in 4-in. and 3-in. diameters. (The vent inlet is 2 in. in both fittings.) This fitting, like the sanitary tee for a sink, can be used only in the vertical position. The inlet to the 90° elbow, which lies horizontal, always connects to the waste pipe from the toilet. The vertical discharge of the 90° elbow conveys the waste to vertical piping or into an inlet of a change-of-direction fitting (90°, 45° or 22½° elbow). This pattern is too deep for joists; I use it when an upper-story plumbing wall has an open stud bay directly in back of the toilet and I am dropping straight down the lower wall with my drain line. I also use it on first floors with generous altitude underneath, so I can use a long-sweep 90 or two 45° elbows to attach to the discharge of the low-heel vent 90 and turn to a horizontal run.

SIXTH PATTERN   Like the fifth pattern, the sixth pattern is used under first floors with generous altitude, where more vertical piping or a long-sweep 90 can be attached to the discharge of a vertical run 4x2 or 3x2 combination wye. I use this pattern so I can swivel the vertical combo inlet branch (and its vent line) in any direction I need to reach a vent wall (which might not be the plumbing wall); I can also swivel the upturned long-sweep 90 in any direction that I want on interior bathrooms and 180° on exterior plumbing walls.

PATTERN 5

Drain line approaches plumbing wall at 90°; vent joins drain at low-heel vent 90.

PATTERN 6

Drain line and vent line can be at any angle; vent joins drain at combo fitting and long-sweep 90°.

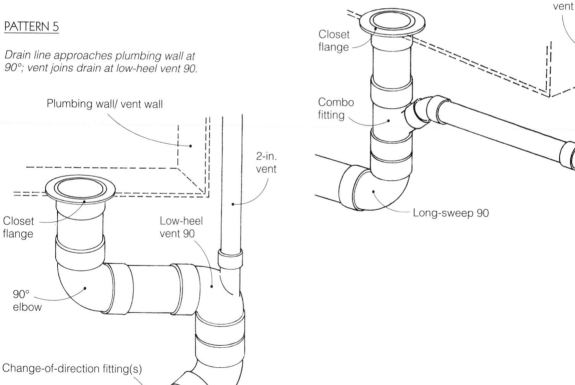

WALL-HUNG TOILET PATTERN

Wall-hung toilets are rarely used in residential construction because they are very expensive, not only in price but also in framing (a 2x4 wall is not strong enough to hold the toilet plumb) and in plumbing labor. If a wall-hung toilet is in the specs, you will need to get the manufacturer's rough-in schematic for the particular model. With a slab foundation (see pp. 131-137), the recommended rough-in height of the fixture might mean that you will need to install the drainage fitting before the pour, because its discharge might need to be below the finish level of the slab. If this is the case, you will need two test caps for gluing to the branch inlet and top inlet to close off the openings before the pour.

BIDET PATTERNS

Bidets are technically classed with toilets because they also convey solid waste, however minute the amount. Twenty years ago, one popular bidet design used a trap under the floor. Today, most bidets have a trap inside the confines of the fixture and use a vertical-run sanitary tee in the wall, just like sink patterns (see below). Unlike the sink pattern, though, the rough-in height of the tee is quite low to the floor. To set your sanitary tee at the proper height, follow the bidet manufacturer's rough-in schematic.

LAVATORY AND KITCHEN-SINK PATTERNS

I wish everything in plumbing were as simple as sink patterns. For sinks we have one fitting: the sanitary tee. As shown in the drawing on p. 106, it is always used with the run in a vertical position, its discharge below the bottom sweep of the branch. The vent always begins at the top inlet and the trap arm at the branch inlet. Because the sink's drain hole is so generously high off the floor, the trap arm can always connect to the branch inlet of a tee. The reason for this uniformity is that every vent (except the vent for an island sink, which is discussed on p. 107) is in the plumbing wall.

On sinks, the length of the tailpiece (the pipe immediately below the fixture's drain hole) is limited by code. My major code wants the vertical distance between a fixture's outlet and the trap weir (the upper flood level of the standing water seal) to be as short as practicable, but not to exceed 24 in. in length. If

BIDET PATTERNS

TRAP UNDER FLOOR

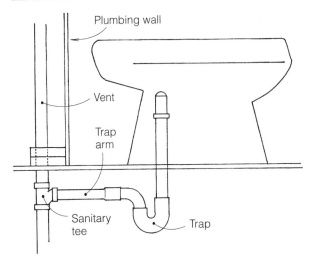

TRAP INSIDE FIXTURE

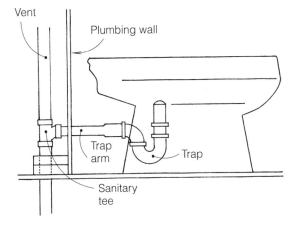

the tailpiece is short, the water draining through it won't be moving fast enough to siphon out the trap's water seal.

The height at which you install the sanitary tee varies with the type (cabinet mounted, pedestal or wall hung) and particular model of sink. For cabinet-mounted sinks I install my sanitary tee branches 14 in. above the finish floor when I have a cleanout above the sanitary tee. When I have a cleanout below the sanitary tee, I place the tee's branch 16 in. off the floor. Various pedestal sinks have different access heights in the rear of the pedestal, and the manufac-

LAVATORY

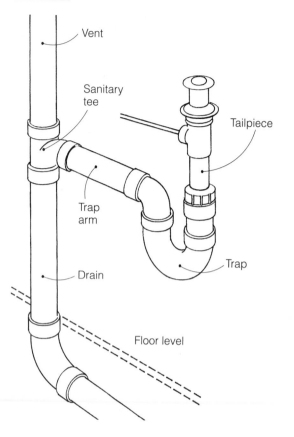

KITCHEN SINK: CLEANOUT ABOVE SANITARY TEE

KITCHEN SINK: CLEANOUT BELOW SANITARY TEE

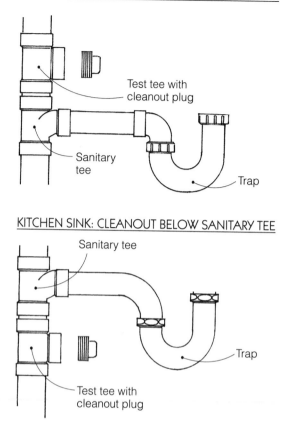

turer will recommend specific heights for the drain and water — check the schematic for the model that will be installed. Since virtually all kitchen sinks are cabinet mounted, I vary the height of the tee or tees for only two factors: depth and number of the bowls and what model (if any) garbage disposer is going to be used. Very deep bowls with the added depth of the basket strainer call for a lower tee. And some models of garbage disposers have an outlet elbow quite low on the appliance. Again, you really should know what make and model of sink and garbage disposer you will be dealing with later on, so you can make provisions now.

For an undersink cleanout, I like to install test tees (which have no sweep and use a threaded plug for sealing) as a means of getting into a drain line. The

test tee may be installed above or below the sanitary tee (see the detail drawings above). I prefer to have the cleanout tee above the sanitary tee on kitchen sinks. A kitchen-sink drain often clogs at the juncture of trap arm and sanitary tee, especially when someone has fed a garbage disposer something it shouldn't eat. If the cleanout is coupled directly to the top inlet of the sanitary tee, the clog can often be removed with a piece of bent coat-hanger wire or a plumber's non-powered snake. On kitchen-sink drains, lower the sanitary tee to about 14 in. off the finish floor and install the cleanout above it, to make sure that access to the threaded plug will not be blocked by the back wall of the sink. On lavatory drains access is not a problem, so you can install the cleanout below the sanitary tee.

LOOP VENT FOR ISLAND SINK

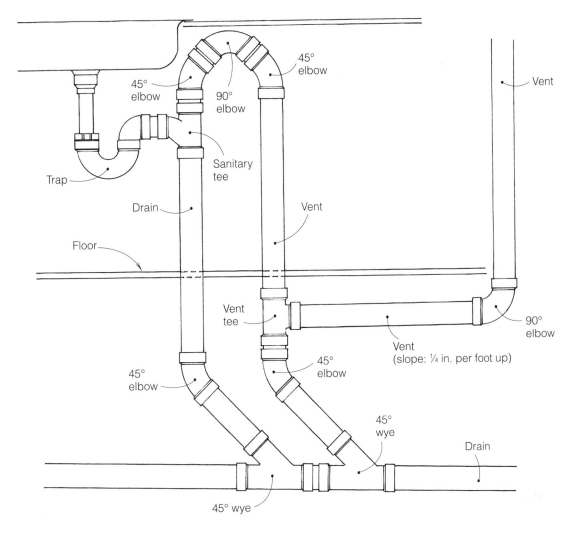

ISLAND-SINK PATTERN Island sinks pose a problem for the plumber because there is no adjacent wall in which to hide a vertical vent run. Instead, part of the vent-to-drain pattern is housed in the sink cabinet, and an extension of the vent portion of the pattern travels under the floor as far as necessary either to be back-vented into another properly sized vertical vent or to be run individually up to the roof. This arrangement is sometimes called a loop vent, and can be used in either a slab or an off-the-ground foundation.

As shown in the drawing above, both lower portions of the loop's legs attach to the drainage line, and the extended vent attaches to a vent tee's branch on the leg of the loop opposite the drainage leg. The top of the loop, a drainage 90° elbow, is positioned as high as possible under the sink. Two 45° elbows start the two parallel loop legs downward from pipe sections out of the 90° elbow. Under the floor, two more 45° elbows and additional lengths of pipe bring both legs of the loop into vertical 45° wye branches on the drainage line. Loop vents can be very expensive in time and materials. Before executing this pattern, ask your local inspector if there are any shortcuts you can take, such as an in-line vent, which in some parts of the country can be used instead of a full loop and exterior vent.

WASHING-MACHINE STANDPIPE PATTERN

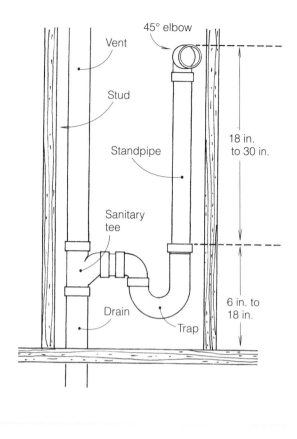

Vent

Stud

Standpipe

45° elbow

18 in. to 30 in.

Sanitary tee

6 in. to 18 in.

Drain

Trap

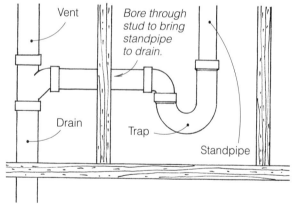

Vent

Bore through stud to bring standpipe to drain.

Drain

Trap

Standpipe

If entire pattern is done in one DWV material, it can be completely inside the wall in one or more stud bays, with only a 45° elbow poking out. If the trap is to be tubular brass, it must be outside of the wall. In that case the branch of the sanitary tee would face out of the wall.

WASHING-MACHINE PATTERN

If your washing machine does not drain into a laundry sink, you will need to install a tall vertical pipe called a standpipe. As shown in the drawing at left, at the bottom of the standpipe are a P-trap and trap arm that join the branch inlet of a sanitary tee. If you are using DWV pipe and fittings and a unionless P-trap, the entire pattern may be installed inside the wall; if you are using a tubular brass P-trap it must be outside the wall. If your standpipe is one or more stud spaces away from the drain, you will have to bore holes through the stud(s) for the trap arm.

My major code does not permit the trap for a washing-machine standpipe receptor to be installed below the floor, and it specifies a height above the floor of not less than 6 in. and not more than 18 in. (This is a safe guideline for lavatory and most kitchen-sink traps as well.)

TUB AND SHOWER PATTERNS

Tubs and showers rest on the floor, so the only place to attach a vent is underneath the fixture. However, structural members may interfere with certain fittings, so you may not be able to use the same pattern for every floor-mounted fixture.

Because of body hair and fats from soaps, the tub and shower drains are prone to stoppages, so you shouldn't skimp on pipe size. Major codes assign a single-stall shower a fixture unit value of 2 and a minimum trap and trap-arm size of 2 in. Bathtubs are also rated at 2 fixture units but are allowed trap and trap arms of 1½ in. because of the holding capacity of the tub. However, I prefer to plumb tubs with a 2-in. trap and drain because they drain faster and stay cleaner. There are two vent-to-drain patterns for shower stalls and tubs.

FIRST PATTERN   The first pattern is used on shallow slabs and sometimes tucked into the joist bays on upper floors. A 2-in. ABS or PVC solvent-weld P-trap accepts either a straight section of 2-in. pipe (riser) down out of a shower drain fitting or the 1½-in. tailpiece from the bottom of the tub waste and overflow with the use of an increasing adapter. A trap that will be inaccessible for service in the finished structure cannot be installed with a union nut (the union nut allows the trap to be opened for cleaning). Instead, you must use a trap that has the trap arm 90 and riser cemented into the J-bend return. Cementing the trap components ensures that they will never separate and cause water damage to the foundation or pose a health hazard due to the escaping liquid waste. As close as possible to the trap arm 90, cement a sanitary tee, with the branch vertical and with the sweep going in the direction of the drainage flow. Now, cement a short nipple and vent 90 elbow or a street 90 elbow to the branch and swing the elbow so that it points to the wall you will be bringing the vent up through. Under this wall, out of the top inlet of the vertical 90° elbow, the vent line will continue upward. The drain line will continue out of the discharge on the horizontal run of the sanitary tee.

SECOND PATTERN   The second pattern, which is the same as the sink pattern, can often be used on a tub or shower. You might use it for a tub where the vent wall is also the valve wall and the drain line needs to drop straight down into a nearby building drain, or for a shower stall where generous altitude allows for vertically positioned change-of-direction fittings. You might also use the second pattern where structural obstructions best left unaltered preclude the use of the first pattern. The second pattern consists of a sanitary tee, with the run vertical. The vent is attached to the top inlet, the trap arm to the branch inlet, and the drain to the bottom discharge. The branch of the tee should be positioned so that there is no more than ¼ in. per foot fall out of the trap arm into the branch. Most major codes want the trap to be within 5 ft. of the vent, for 2-in. traps and pipe. Using a 2x1½x2 sanitary tee allows you to install 1½-in. pipe for the vent.

TUB/SHOWER PATTERNS

PATTERN 1

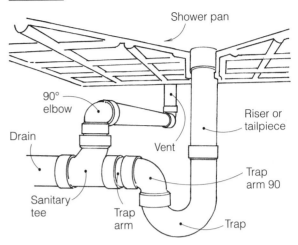

PATTERN 2

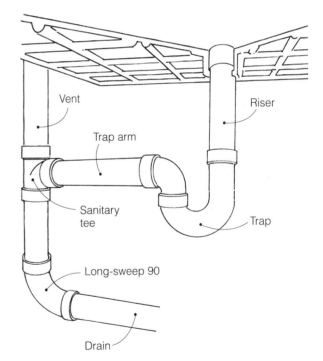

CONTINUING VENTS UPWARD

In DWV plumbing, the configuration of the vents plays a critical role in how well the system will function. The speed and quantity of the air that flows through the vent piping affects how well your toilets flush and how fast your bathtub drains. Just as you want the effluent to flow as smoothly as possible in the drainage fittings, you want the outside air to flow as smoothly as possible through the vent fittings. As wastes are sent down the drainage system, air is pulled in through the vents. If fitting sweeps are positioned to follow the flow of incoming air, the system will "breathe" properly. If the sweeps are not installed to follow the flow, the increased friction will slow down the drain rates.

Every plumbing fixture needs a vent, and the simplest way to visualize the vent system is with each vent running straight upward, through the roof. However, that vent configuration is not desirable, since it's best to minimize the number of holes in the roof, and sometimes vents can't run straight up because they need to bend around obstacles, such as other mechanical lines or cabinets. Holes in the roof are kept to a minimum by joining individual fixture vents before exiting the roof; this back-venting, as it is called, saves on vent materials as well.

BENDING VENTS AROUND OBSTACLES

In my major code, each vent is supposed to rise vertically not less than 6 in. above the flood-level rim of the fixture served before offsetting horizontally. Aside from having a vent run of the proper diameter for the number of fixture units it's venting, this is the next most important vent regulation you will encounter in residential work. In the case of a lavatory basin that will probably have a rim height off the floor between 31 in. and 33 in., the vent must be 39 in. from the floor before it can go sideways.

What happens if the 6-in. rise requirement is in conflict with an inset medicine cabinet above the lavatory? Your local inspector might allow the vent to run one stud space over from the drain (see the drawing on the facing page). In this situation you need to get out your boring bits and drill motor.

Choose the bit sized for 1½-in. pipe (in my Milwaukee drill index, it's the 2⅛-in. self-feed #48-25-2120). When drilling through studs and then up through plates, always try to avoid having to drill through nails. You might also have to avoid sheet-metal ducts and wiring. You will probably be working before the electrician or HVAC installer, so check the plans. If the ducts are already installed, you have to vent around them. Depending upon what you find in the walls, you might have to run the vent horizontally more than one stud space before turning upward.

Begin by boring a hole through the lower plates of the second story, then bore another hole through the first floor's fire blocks. You can use a plumb bob to align the hole in the plates to mark the next hole in the block below.

At the lower end of the run, make sure you have a stub sticking out of the top of the sanitary tee at least 6 in. above the flood rim of the basin-to-be. Get a vent 90 elbow and hold it up next to the stub so that the socket is at the level where it would be when cemented and shoved onto the stub as far as it will go. (As discussed on p. 54, you can't push the elbow's female opening onto the male stub dry without a lot of force, and lubricants would prevent a good cement bond.) With the elbow held in place with one hand, measure the height (from the lower plate) with the other hand, to the center of the horizontal socket of the elbow. Make a mark at this height on the stud to be bored. Then bore through the stud. You should still be well below any fire blocking.

If the fire blocking does interfere with the up-turned 90° elbow, bore a 2⁹⁄₁₆-in. hole in the block, which will be wide enough to accommodate the hub of the top socket on the upturned elbow. (If you bore a 2⅛-in. hole first and then discover that the 90° elbow needs to be right inside the block, you won't be able to enlarge the hole accurately, and in most cases you will split out the block trying.)

Sometimes you need to run more than one vent in a stud bay. With more than two vents, you can't usually bore upper plates and blocks solely with boring bits. You'll be hitting nails and need to use the hole saw on some or all of the holes. To get three or more vents in one stud bay and have room to back-

VENTING AROUND A MEDICINE CABINET

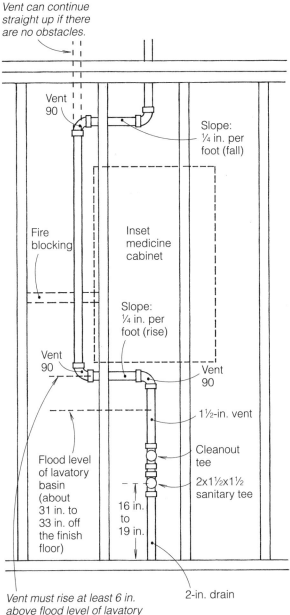

Vent can continue straight up if there are no obstacles.

Vent 90

Slope: ¼ in. per foot (fall)

Fire blocking

Inset medicine cabinet

Slope: ¼ in. per foot (rise)

Vent 90

Vent 90

1½-in. vent

Flood level of lavatory basin (about 31 in. to 33 in. off the finish floor)

Cleanout tee

2x1½x1½ sanitary tee

16 in. to 19 in.

Vent must rise at least 6 in. above flood level of lavatory basin before running horizontally.

2-in. drain

vent them on this floor or the next, you must have a vent tight against each stud of the bay with a third in between and placed in the bay on spacing that allows for the fittings required for back venting (see pp. 112-113).

BACK VENTING

Whenever possible, I try to combine as many vents into one before they go out the roof. This reduces the number of openings (and possible weather leaks), and on many structures greatly improves the appearance of the roofline. It can also save labor time and material costs. Back venting in the attic is usually simple enough because there's room for you to move around in and plenty of open space around the vents.

Back venting in a wall, however, can be a tedious task. This situation often arises with a toilet and a lavatory on the same wall; you will want to back the lavatory vent into the toilet vent. First, check your local code for this juncture height, which might differ in various parts of the country. My major code says 42 in., but some inspectors want the juncture 54 in. from the finish floor. At whatever height you are instructed, you can place a vent tee (in plastic pipe up to 3 in. in diameter) with the center of the branch inlet at this height in the toilet vent. In the case of a 2-in. toilet vent, it would be a 2x2x1½ reducing vent tee (see pp. 18-19). For a 3-in. toilet vent on a main stack, you would need a 3x2 bushing for the tee's branch since there is no 3x2x1½ reducing vent tee. You would put a vent 90 elbow (or drainage 90 elbow) on the lavatory vent at the prescribed height and run your 1½-in. vent over and into the branch of the vent tee. If your local supplier does not have any vent tees, you can use a sanitary tee instead, but you must invert it so that the pitch of the branch sweeps downward. Bore your holes through the necessary number of studs so that the vent from the lavatory sink coming over to the toilet vent has a ¼-in. per foot increase in altitude.

Sometimes it is inconvenient to bring the two vents together right away in the stud wall, so you have to find another solution. If there are too many obstacles in the stud wall between the toilet and lavatory and your structure is single story with an attic and you are not on an exterior wall, you can poke through the upper plates with the two vents. As close to the tops of the ceiling joists as possible, you can add a 90° elbow to the lavatory vent and run over to a vertical tee (still vertical) on the toilet vent. If you are on an exterior wall and your ceiling joists are running in the direction you want to go with the vents, you can have both lines horizontal after going

BACK-VENTING OPTIONS

IN THE WALL

Vents are joined in vertical plane.

IN THE ATTIC

Two vents are joined in horizontal plane.

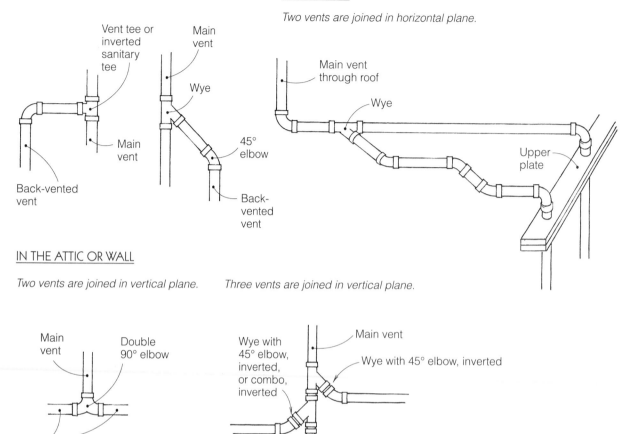

IN THE ATTIC OR WALL

Two vents are joined in vertical plane.

Three vents are joined in vertical plane.

through 90° elbows and then use two 45° elbows with a sufficient length of pipe between them to get up above the ceiling joists; then join the vent runs on the flat with a wye or combination wye and one more 45° elbow. If you keep this juncture as low as possible, you still might have the necessary altitude to back this combined vent and any other vents into one 4-in. vent before going through the roof.

On attic spaces with plenty of unobstructed square footage but not so much altitude, you can plug the flat-run 1½-in. and 2-in. vents into branches of flat 4x2 wyes coupled with 4-in. pipe in the barrels, provided that you have the wyes' branches swept back

toward the direction of the fixture. If you have the altitude, you can stack inverted (branches sweeping downward) 4x2 combination wyes and come at the big 4-in. vertical vent like spokes of a wheel centering on its hub.

Joining vents in the wall is always more tedious than joining them in the attic because of restricted space and the need to bore holes, but sometimes you have to do it. To join two or more vents in a vertical plane, you can use a double elbow or stack wyes and combos. These patterns can be used either in the attic or in the wall.

ASSEMBLING THE VENT

Once you've got aligned holes through the fire blocking and upper plates and maybe even the roof, you're ready to install the roof vent, through which air will enter the drainage system. If the house is a one-story structure, your local authority may permit ABS or PVC vents above the roof. Many communities won't, and you will have to adapt to either DWV copper or no-hub iron before the vent exits through the roof. If you can have plastic above the roof, then measure the distance of the intended pipe run from the top of the hole bored through the fire blocking to a point 1 ft. above the roof. If no plastic is allowed out the roof (some codes bar its exterior use because sunlight can make it brittle), measure the distance 2 in. above the upper plates to the top of the hole through the stud.

Now you can take your reciprocating saw and cut the nails holding the fire blocking in place. Then cut a piece of pipe of the appropriate length. Slide the block over the pipe and then shove the pipe up through the hole in the upper plates; toenail the block back into position or use the cordless and screw it back in place. Now shove the upturned 90° elbow onto the new vertical vent run just hard enough so that it stays there (A in the drawing at right). Lift up the vent pipe until the 90° elbow aligns with the hole in the stud and measure the distance between the two elbows, allowing for full penetration into each female opening. Cut a section of pipe this length (B). Slide it through the hole in the stud and apply cement to both pipe and the 90° elbows and bond them together. Next, cement the bottom discharge leg of the 90° elbow on the right to the end of the short vertical length of pipe (C). Now lift the long vertical vent run up and apply cement to it and in the upturned elbow's female opening and then drop the vent down and mate the two parts. If you do not want to have to use a Mission coupling or other locally acceptable band seal as a union, you have to cement your pipe joints by working in one direction only and always making sure that the two parts to be cemented can be drawn apart and then together. Sometimes this is impossible and you must use the band-seal type coupling as a union to join two butting ends of pipe.

ASSEMBLING A VENT

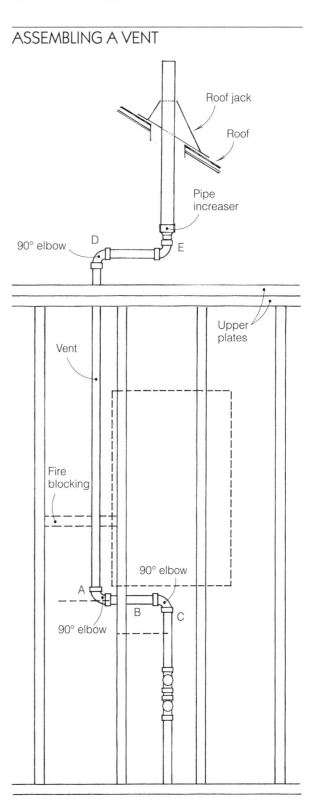

On exterior plumbing walls (they are allowed in many areas), you may not want to take the vent straight through the roof in line with the wall because it could weaken the structure and/or the vent's flashing will be too close to the edge of the roof for a good flashing/roofing job. If there are ceiling joists and a drywall ceiling, you can use two more 90° elbows to offset the vent pipe far enough away from the edge of the roof for a proper installation (D in the drawing on p. 113). If you have an operable skylight, the vent must be above it and 10 ft. away from it.

In my locale, it is not a practice to install hoods or 180° returns on the upper terminus of the vent (although this is allowed elsewhere). My major code states that in areas where snow and frost may clog the vent, each vent extension through a roof shall be at least 3 in. in diameter, and that the change in diameter shall be made inside the building at least 1 ft. below the roof and terminate not less than 10 in. above the roof. (Check with your inspector for a similar statute.) If you are running a 1½-in. or 2-in. vent, you will need more than 1 ft. of altitude in the attic to adapt from the smaller pipe to the 3-in. pipe.

When you bring the 1½-in. or 2-in. vent through the upper plates, you can make it sufficiently tall so that you can add a vent 90 elbow and swing fit (rotate it in any direction) just above the tops of the ceiling joists if it can't go straight up. Then add enough pipe in the elbow's horizontal opening to move to an area with the most generous altitude before adding an upturned vent 90 elbow. Off the top of the upturned 90 elbow, stub in a short piece of pipe and then cement on a 1½x3 or 2x3 pipe increaser (E). Cement the smaller fitting's hub to the short stub and then cement a piece of 3-in. dia. pipe into the top hub and go out the roof. Make the short stub as high as you can, while still leaving at least 12 in. (below the roof sheathing) of 3-in. pipe going out the roof.

INSTALLING A ROOF JACK   The roof jack seals the opening around the vent pipe and helps to support it in a vertical position above the roof. Depending on the job, it may be installed by either the plumber or the roofer.

The standard roof jack has a lopsided metal cone. The steep-angled side of the cone should face the ridge; the long side slopes in the direction of drainage. There is also a roof jack with a rubber cone piece

ROOF-JACK INSTALLATION

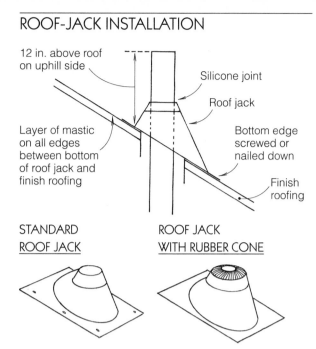

12 in. above roof on uphill side

Silicone joint

Roof jack

Layer of mastic on all edges between bottom of roof jack and finish roofing

Bottom edge screwed or nailed down

Finish roofing

STANDARD ROOF JACK

ROOF JACK WITH RUBBER CONE

that grips the circumference of the vent when it is slid down into place. I like the rubber-cone version better but it needs protection with aluminum paint.

The roof jack is fastened to the roof with screws or nails through holes that you punch along the sides and sometimes the top and bottom edges. On steeply pitched roofs it's better not to punch holes on the bottom edge because water tends to hang up on the screw or nail heads and rust them, creating the potential for a leak. The steep pitch should be enough to ensure fast rain runoff. Flat roofs need holes on all four edges. On a wood- or asphalt-shingle roof, the top edge of the roof jack should slip up under a course of shingles, if possible. On the uphill side, no holes are necessary. on the downhill side, predrill holes in the shingles.

On most roofs, all the underside edges of the jack should be generously coated with roof mastic before the jack is slid down the vent stack onto the roof. Then start your screws or nails in the center of the sides and work out to the corners. I strongly suggest that you use stainless-steel screws or nails for this application if you can find them. If you are using a standard roof jack, seal around the top edge and the protruding vent with silicone sealant. I coat the sealant with standard black mastic when it is dry and spray-paint it with aluminum or galvanizing paint. If you are using a jack with a rubber cone, protect the rubber from sunlight by coating it with roof mastic (which I apply with a gloved hand); then spray paint with aluminum paint.

DWV PLUMBING WITH AN OFF-THE-GROUND FOUNDATION

Let's now consider a typical DWV installation in off-the-ground construction, that is, a house built on a crawl space or basement. Assuming that the house has concrete perimeter foundation walls and a grade beam or two within the inside limits of the foundation, there needs to be a way for the drains (and sometimes gas and water lines as well) to pass through the walls. If the foundation crew doesn't do it, you may need to install through sleeves in foundation forms before the foundation is poured.

How you proceed with the rest of the work depends in large part on how accessible the crawl space or basement will be after the structure is completed. For structures that will have a workable crawl-space height, you may not need to get involved any further with plumbing until the stud wall is up. For structures with inaccessible space between the bottom of the floor joists and grade, you have to plumb the building drain as for a slab, before the floor is installed, by laying the drain, waste and vent pipes and then only stubbing up fixture drains and any vents to above the plates (see the discussion on pp. 131-137). For structures with expansive first-floor plans, you might want to start plumbing as soon as the floor joists have been set but before the subfloor is laid. (It helps a lot to be able to stand up between floor joists when hauling in full lengths of pipe.) If you can accurately determine wall positions without the floor being down (see the drawing on p. 133), you can go ahead and hang the building drain from the floor joists now, leaving 2 ft. extra of plumber's tape so you can fine-tune the fall of the drain later on.

You might not have enough time to hang more than one pipe system before the framers want to get on with their subflooring. Usually they can be convinced to lay the floor on those areas devoid of plumbing first, and then come back and cover up the piping when they run out of other things to do. So if any piping can be or needs to be hung from the first-floor joists at this time, it should be the larger drains. Smaller-diameter, somewhat flexible, lighter-weight drain-line material can be placed under the structure after the subfloor is down. If the job foreman agrees, get as much piping as possible installed or at least dropped in the crawl space before the subflooring is nailed down, even if you just hang rough lengths temporarily from the joists in the general direction of their future installation. Working this way is easier on your back and will save you time later.

On two-story structures, I think no-hub iron is the best material to use for upper-story toilet drains and their drops. I often plumb below the first floor in ABS or PVC, with the remaining portions of the first and second floor in iron or copper. However, of late, some inspectors have been demanding that I plumb structures of only one material, even though the mixing of materials on two stories is not prohibited by my major code. This is not to say that you should never do a two-story structure totally in ABS or PVC; just remember that plastic toilet drains tend to be noisy.

Various straps and hangers are available for hanging different piping materials (see pp. 34-38). Plastic pipe poses little problem since it is light in weight and easy to maneuver. Iron pipe is a bit more difficult to hang because it is so heavy (see the sidebar on pp. 116-118).

UPPER-STORY PLUMBING

I start my toilet DWV on the uppermost floor first, to beat the electrician and sheet-metal installer to the most usable stud bays. (In basements and crawl spaces, there is always enough room to go around.) On second and higher floors, floor joists largely determine the pathways for the piping, and their depth can determine the diameter of pipe and fittings as well as the vent-to-drain patterns, especially for toilets (see pp. 101-105). The shallowest floor joists you will normally encounter are 2x8s, with 2x10s or 2x12s more likely; 2x14s are a rare luxury.

When you encounter 2x8s on an upper story (where I install no-hub iron), because of the radius size of a 90° closet bend and because of the need for the drain to slope slightly downward, you have to use 3-in. pipe. Even with 2x10 joists, if your horizontal run to the point of drop is more than about 8 ft., this

It is important that vertical and horizontal DWV iron pipes be well supported, both for the integrity of the system and for your own safety. For many installers, this is an area of little glamour, and if a job is left wanting it is often on this issue.

VERTICAL PIPE RUNS

My major code calls for supporting vertical cast-iron pipe at each story or closer. This can be done with riser clamps (see p. 34) that bolt to the pipe or fitting at the lower plate, if there is enough room. This clamp is a very crude design but it works well to support heavy material. However, on 2-in. waste lines for lavatory, kitchen-sink, bathtub and laundry standpipe, the accompanying copper fresh-water distribution piping often interferes and precludes the use of this clamp.

When I can't use a riser clamp when running vertical 2-in. iron lavatory, kitchen-sink and laundry waste lines, I use galvanized plumber's tape to support the sanitary tee. Even with a bolted-on riser clamp at the lower plate, the weight of the vent line above can still cause the branch of the tee to sag or point downward due to the flexibility of the no-hub couplings. Screwing the tape to the inside of the studs about 4 in. or so above the top of the tee and cradling the branch will take care of the problem.

If there are 45° offsets in a vertical no-hub iron vent line in a wall, the diagonal section of pipe may need support, depending upon how long it is and how much vent lies above it. The easiest way to support diagonal pipe is to take a 6-ft. strip of galvanized tape and divide the length in two. At the center of the tape, wrap it around the pipe from below, right at the bottom of the upper 45° elbow, and put a bolt through the two closest matched holes. Run a nut up and snug it down tight. Then in a straight line (45° to vertical) carry each leg of the tape up to each side of the stud, and while taking the weight off the diagonal run, nail or screw the tape in place.

HORIZONTAL PIPE RUNS

In horizontal piping, there are more options because of the presence of floor and ceiling joists (see the drawing on the facing page). I advise supporting horizontal runs every 4 ft., within a few inches on both sides of a coupling or a branched fitting, and at the first coupling of a branch and intersecting pipe run, whenever it is possible.

If the subfloor is already down on an upper story, to hang the drains you usually have to work from underneath, on ladders. The biggest problem when working under a floor is your impeded ability to screw the plumber's tape to the insides of the joists. I have found that 2x4 blocks sized to span the joist bays can be toe-nailed or toe-screwed to support the drains (4 ft. on center) after you have finished hanging the runs temporarily with plumber's tape (see the drawing below). When the drains are laid in for good, you can wrap a strip around the drain and screw or nail it to the edge of the block to prevent the drain line from moving laterally out of its intended path.

In the basement or crawl space, if the pipe runs perpendicular to the joists, I might first use a chalkline and snap the run on

SUPPORTING HORIZONTAL PIPE FROM BELOW

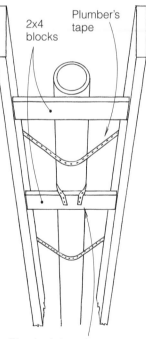

2x4 blocks

Plumber's tape

Plumber's tape here prevents lateral movement of pipe.

HANGING HORIZONTAL IRON PIPES

PIPE PERPENDICULAR TO JOISTS

Joist

Pipe

Tape

PIPE PARALLEL TO JOISTS

Joists

Tape

Pipe

PIPE DIAGONAL TO JOISTS

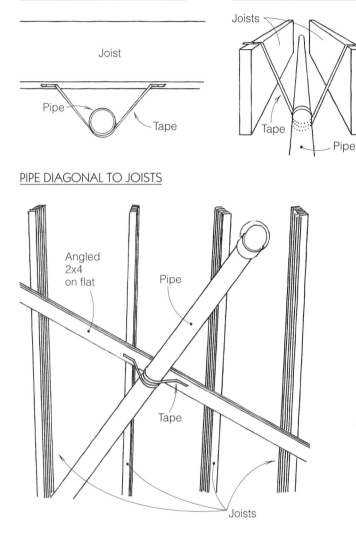

Angled
2x4
on flat

Pipe

Tape

Joists

the bottom edges of the joists. Then, using a cordless screw gun, I screw one end of an overlong length of plumber's tape to the top center of one joist. (Use two screws or nails each time you secure your plumber's tape when hanging iron systems. Come in one or two holes from the end. Then go down under the fitting — 2 in. back from the edge — and then around it and back up to the top edge of the other joist.)

I place my screws about 2 in. apart, using the smaller-sized holes if the tape has an alternate-hole pattern. If you use the large hole and turn the screw too tightly, the head of your fastener won't cover up the entire hole. and the weight of the pipe will be carried just by the side edges of the tape at the hole. The tape might tear and break under the stress. If you use the small hole, the screw covers the entire hole, so more of the width of the tape carries the weight.

If the only tape you can get has the larger, same-size hole pattern, use screws with a head large enough to cover the hole or long, wide-headed roofing nails that are used on roofing felt. Try to cut the plumber's tape as square as you can. Whether you use nails or screws, leave enough tape sticking up above them so that you can fold it down over the fastener heads later. That will make your work look a lot neater.

If you are working alone, hanging horizontal iron drains can be a challenge because of the weight of the materials. I lift up the pipe on one end and wrap the tape around the pipe so I can just get it off the ground on that end and screw it to the joists. This holds one end of the pipe. Then I go to the other end and do the same thing. Because fine tuning will probably be necessary and because the iron pipe and fittings are so heavy, I leave at least 1 ft. of galvanized tape flying in the wind (2 ft. is better) after securing it to the second joist. That is enough length to be grabbed by two gloved hands comfortably for lifting and lowering the pipe run with sufficient control. Once final adjustments have been made, the excess tape can be trimmed off. The countersunk heads of the screws pull the galvanized tape into the top edge of the joist far enough so that there is no lump to affect the flooring.

On long runs, when working by myself, I put a sling of plumber's tape at each end of the run, about 1 ft. in, and set the slings lower than what I think the eventual elevation of the pipe will be. I can then lay the pipe in on top of the slings and work it up and down to adjust its fall. Then I add permanent supports on 4-ft. centers, making my first tape wrap all the way around the pipe with alternates slung under and pinched with bolts as close to the top of the pipe as possible. After all the necessary supports are in place, I either remove the slings or make them into full wrapped supports.

When there are change-of-direction fittings in the line (and nearby structural members), I hang the pipe within several inches of a coupling. Many times this means adding a 2x4 block for attaching the tape to if I'm between joists at this point. I like to use 3½-in. grabber-type screws (these have steeply angled threads and larger heads than drywall screws) to install the block. You can usually use your shoulder (or your knees if you're working in a crawl space) to lift and lower the pipe to the right height while you tighten the fasteners. I generally hang the system with just enough tape to see its final configuration. When I'm satisfied with the arrangement, I go back and add additional tape on 4-ft. centers and put additional strapping at couplings and fittings.

If the pipe runs parallel to the joists, instead of adding blocks every 4 ft., I cradle the pipe by screwing the tape to one joist, then bringing it down and around the pipe once, then going back up to the opposite joist and screwing it to that one. Here again I let the tape run wild for now. If most of your hanging tape ends up being one vertical strand with nut and bolt, some inspectors like to see you add a triangulated support (cradle) every so often to reduce the possibility of any side sway. On runs perpendicular to the joists, you just cradle to a single joist.

If the pipe runs diagonally to the joists, a likely event where it joins the building drain in the basement or crawl space, you may have to nail, screw or bolt a 2x4 on flat (and angled perpendicular to the pipe) to the bottoms of the joists to give you something to anchor your plumber's tape to (see the bottom drawing on p. 117). My major code says that where joints occur, suspended iron pipe should be supported at no less than 5-ft. intervals. If you come down off the joists and wrap the plumber's tape around the horizontal piping, keeping the tape flat on the pipe and in a vertical plane, then the piping will remain in position, even in mild seismic activity. (If the tape is twisted and the two support legs are not in a straight plane, the pipe tends to move around, especially in earthquakes.) You can create a doubly effective check on horizontal movement by bringing the tape together at the top of the pipe after cradling and installing a nut and bolt through opposing holes as close to the top of the pipe as you can. Do this at alternating support locations.

added need of fall (⅛ in. to ¼ in. per foot) might mandate 3-in. pipe and fittings in order to stay within the height of the floor joist. If I have to use 3-in. pipe, I always use a 3x4 increaser and convert the drain to 4-in. pipe as soon as possible, usually on the drop, within the first several feet below the upper plate, on 2x6 walls. If your wall is all 2x4 to the crawl space, you'll have no other choice but to run the entire drop in 3-in. pipe. At the bottom, you can use an increasing coupling or a bushing and a 4-in. 90° elbow and continue horizontally in 4-in. pipe.

If you are dropping 3-in. iron drain down through an interior, non-bearing 2x4 wall, remember that a bored hole for 3-in. pipe effectively eliminates the total width of a 2x4. You might as well save yourself a lot of time and sweat and use the reciprocating saw to cut through the plates. Once the plates are cut away, you'll have to nail or screw heavy metal reinforcing straps (generically called FHA straps) across the cut plates on both sides. No one in his right mind would bore (or saw) upper and lower plates for a 3-in. pipe drop in an exterior 2x4 wall. That isn't to say that it isn't done or that an inspector might not look the other way and pass such a mistake. So if you have yet to design or have your future structure or remodel designed, I suggest that you designate as many interior walls as possible as plumbing walls (in cold climates, plumbing walls are required by code to be on the interior of the building).

PLUMBING A TOILET   I always try to use no-hub iron for second-floor drains because it is much quieter than plastic piping. I will describe how to install no-hub in this section. If you choose to plumb with plastic, you can follow the same sequence, and your job will be easier.

On a second floor, begin rough plumbing with the toilet drain that is farthest upstream. If the subfloor is down, you'll have to cut a hole for the closet flange. First, lay out the centerline of the drain on the subfloor and lower plate, following the recommendations of your plumbing code. Let's assume you are installing a 12-in. rough toilet (which requires 12½ in. to 13 in. of clearance to the plumbing wall). Most major codes insist on a minimum of 30 in. between "similar" fixtures (centerline to centerline), an important restriction if you are also installing a bidet. For only a toilet, make sure you have 15 in. plus the thickness of any drywall from the centerline of your bowl to any side wall, cabinet side or bathtub.

TOILET ROUGH-IN DIMENSIONS

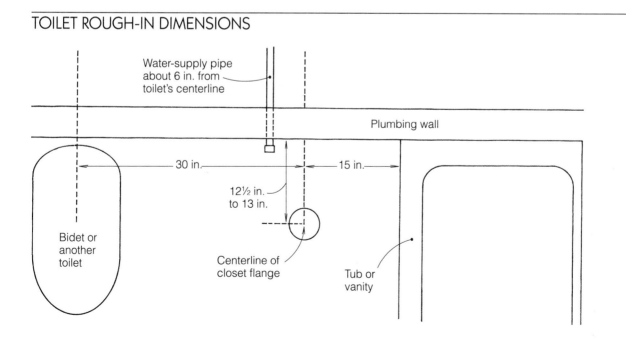

At this point you make a mark on the floor between 10½ in. (for a 10-in. rough toilet) and 14½ in. (for a 14-in. rough toilet) out from the plate, depending upon your choice of bowl size. (The extra ½ in. is for ½-in. drywall. If your finish wall will be thicker, you need to increase this measurement accordingly. It doesn't hurt to allow a little extra space if possible.) Drill a small hole through the floor on your mark. Now go downstairs and see where this hole is in relation to floor joists and blocking. (If you have a single-story structure, then you and your flashlight must go under the house.) All too often I find that my drill bit never makes it through the subfloor into plain view because I am drilling dead center on a joist. Drilling within 2 in. of a joist is an equally bad outcome. When this happens, a carpenter has to do redo the framing, heading off the joist so the drain hole can go where it needs to.

The next step is enlarging the drilled hole to receive the closet flange. Closet flanges are one- or two-piece fittings that anchor the toilet bowl to the subfloor. They attach either to a closet elbow, another change-of-direction fitting or a branch fitting directly below the toilet bowl's discharge or to a short section of pipe that connects to these fittings. I work with two brands of closet flanges: Casper's Industries and Frank's Pattern; each requires a slightly different installation procedure.

The Casper's flange #165 no longer comes with a paper template for laying out the hole it will fit into, so you must make your own. With the aid of a simple compass, I made an acceptable one on my second try in about 20 minutes. You can also use dividers (I made my own set out of flat steel stock) to take the diameter of the flange's hub and transfer it to the floor with the school compass. Center your template over the pilot hole and scribe the floor. Then bore a larger hole on the inside edge of the scribed perimeter and carefully jigsaw the rough opening to make the closest fit you can to the flange. When I cut the hole, I take just a blade width inside the line. Sometimes the cut needs a touchup with the rasp, but never very much.

As seen in the top photo, the Casper's flange has four slots for closet bolts. Two of them are arced with wider openings at the beginning to allow the introduction of the flattened head of the brass machine-thread closet bolt. The other two slots are simple, straight notches with relieved undersides to cradle

At top, a Casper's Industries flange screwed to the subfloor; above, a Frank's Pattern flange being installed. (Photos by Bill Dane)

the heads of the closet bolts. Because the cradling ensures installed bolts that are closer to plumb, I use these straight notches for my installations.

The Frank's Pattern #40 flange (see the photo above) needs no template; you can remove the bottom half of the flange by taking out the three stainless-steel drawbolts and use it to trace its shape on the subfloor. (Before you separate the two pieces, use a crayon or felt-tip pen and mark across both pieces so that you can rejoin them in the same orientation.) Notice that this flange's bottom half has three lugs or bosses hosting tapped holes for the drawbolts. Flip this piece onto the side with the largest diameter (where the lugs contact the floor), so you can get the most accurate scribe. Position the bottom half on the floor so that when the halves are bolted together the two straight closet-bolt notches on the top half of the flange will be 180° to the rough opening you will cut. This means the lugs will be approximately at either 2 and 6 and 10 o'clock or at 4, 8 and 12 o'clock.

Neither of these closet flanges has a very wide top flange so there's not a lot of overlap between the flange and the subfloor. To anchor the flange more securely, I drill four to six outwardly angled holes in the outer edge of the Frank's Pattern flange and then sink long brass or stainless-steel flat-head wood screws into the subfloor. It is very important to countersink the holes in the flange's edge, because if you don't, when you snug up your screws you can crack off the edge of your flange and you'll have to buy another one. The Caspers flange has factory countersunk holes but they are real close to the bolt bosses, so if you want to use them you have to be very accurate when roughing in the opening in the subfloor.

The first fitting out of the closet flange is the closet elbow. No-hub 90° closet elbows come in various lengths and heights. On 4-in. drain lines, you can use a 4-in. ¼ bend (the tightest radius in an equal dimension drainage 90° elbow) right out of the closet flange, but the closet bend has a tighter radius than any other possible choice of drainage 90° elbow. If the drain line is only 3 in., the next piece to be installed is a 4x3 90° reducing closet elbow. The top (4-in.) half of the elbow slips into the bottom of the flange. I believe that this bend, with its larger diameter entrance, makes for a better-performing toilet.

On upper stories, where the plumbing needs to remain within the height of the floor joists, you might as well purchase the smallest closet bend. With the longer and taller bends (as they increase in length, they usually increase also in height), you often end up snap cutting off some of the height. Notice the grooves cast into your closet bend near the top of the vertical inlet portion. Think of these grooves as fault lines. They are there to aid you in cutting off just these very small portions of the fitting. You place the wheels of your snap cutter's chain in the groove that best suits your requirements (see pp. 88-89), and the fitting should snap just there, without cracking the entire piece.

From the end of the elbow's horizontal barrel, your next move depends upon which way the floor joists run. You need to choose and install a vent-to-drain pattern (see pp. 101-105) that suits the location. If the floor joists run perpendicular to the wall behind the toilet and there is another wall directly below this one in which the pipe can drop (see the side-

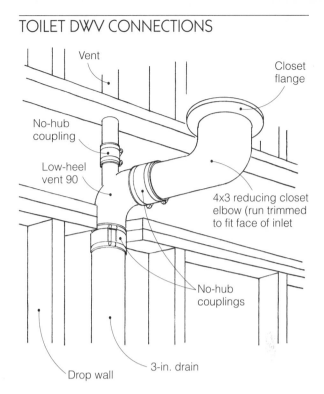

TOILET DWV CONNECTIONS

bar on p. 126), the fifth pattern is a good choice. If the floor joists run parallel to the wall behind the toilet, the third pattern is a good choice.

You need to bore a hole in the plate for the vertical 2-in. vent pipe. Center your mark in the plate (a framing square works well for extending the line from the center of the closet flange) and then use the right-angle drill with a boring bit for 2-in. pipe (probably a 2⁹⁄₁₆-in. bit). In order to get the vertical leg of the low-heel vent 90 in place, you'll need to cut completely through the upper plate of the 2x4 wall below, as shown in the drawing above.

Depending upon the height of the floor joists, the bottom edge of the low-heel vent 90 might end up within the depth of the plate or plates. You then will need to cut out enough plate to get the fitting in with a no-hub coupling on the end. Remember that on a 2x4 wall, the cut will have to be wide enough to allow the no-hub coupling's drive screws to be positioned on one side or the other of the fitting (inside the wall) not in front or back. Otherwise the drywall will be held off the studs. On a 2x6 wall, there won't be a problem.

The heads of the draw bolts in a closet flange are very difficult to tighten without the use of a socket wrench. For now though, you just want to snug up the flange onto the vertical leg of the closet bend. The top edge of the closet bend should be flush with the flat brim of the flange. Tighten the bolts enough to hold the elbow in place; when you are no longer able to rotate the flange on the elbow, it is tight enough. You should now be able to send the lower end of the closet bend down into the hole in the subfloor until the flange is resting in place on the floor. However, the off-center weight of the closet bend will lift the front edge of the flange off the floor somewhat. To remedy this, go below the floor and temporarily hang the closet bend level with some plumber's tape and drywall screws.

Now turn your attention to the low-heel vent 90. With a short piece of 2-in. iron pipe coupled to the 2-in. vent inlet, lift the fitting up, sending the 2-in. pipe up about 3 in. through the hole in the lower plate and subfloor. Lift the fitting until its upper, horizontal leg lies alongside the closet bend's horizontal leg. A plumber's torpedo level has a magnetic strip on one side, which you can use to attach the level to the low-heel vent 90 to make certain that you are holding it near plumb. On the closet bend, mark the edge of the low-heel vent fitting's opening. Now go back topside to pull up the closet bend and snap-cut it. The Wheeler #527 is a good tool to use (see pp. 88-89). Because snap cuts are not flush and due to the need for some extra room for the thickness of the stop in the no-hub coupling's rubber collar, you shorten your measurement by ¼ in.

With the closet bend cut to length, reposition it in the hole with the flange still in good alignment. Put just two screws in the flange right now, at 9 o'clock and 3 o'clock, to keep the closet-bolt slots in proper position. Now go back under the floor, lift up the low-heel vent 90 and 2-in. pipe once more, and when your torpedo tells you that both fitting legs are level, couple them with a 3-in. no-hub coupling. Normally you should have a slope of ¼ in. per foot on the closet elbow, but if you do, sometimes it causes an improper fit between the rubber compression ring in the flange and the bend. If you level both fittings to be joined, the rubber ring will seat properly. Since this pipe run is just 12 in. to 14 in., slope is insignificant.

With the two fittings coupled, check to make sure that the slots in the closet flange are where you want them; then finish screwing it to the subfloor. Then finish torquing the draw bolts on the closet flange.

From the bottom of the vertical discharge leg of the low-heel vent fitting, you use another coupling and as much pipe as necessary to get down to the basement or crawl space. If this drop remains inside a 2x4 wall, you'll be eliminating the upper and lower plates where the drain passes through. If the wall changes to 2x6 under the subfloor, you can use a 3x4 no-hub increaser or a 3x4 no-hub coupling and continue your drop in 4-in. pipe.

Off the top of the 2-in. vent stub that protrudes through the lower plate, you use another 2-in. no-hub coupling and join to more pipe. Regardless of which vent-to-drain pattern you use under the floor, running the vent above the plate will be the same.

When plumbing one-story structures, I try to run each fixture's drain to the building drain individually because the distances are not of great consequence. On a multi-story structure, if only one wall is available for dropping drains, cutting away the wood structure for two additional drains for a bathroom will considerably weaken the wall. In this case, I try to splice drains together before going through the wall's lower plate. This is done by adding wyes in the toilet drain and bringing the smaller drains over to them. If I can get the additional fixture's smaller drain to drop into the same stud bay as the closet drop (by boring studs instead of joists) I then first try to use a 3x2 or 4x2 wye with a 2-in. ⅛ bend to bring the wye's branch up vertically. If I cannot get the smaller drains moved over in the wall above the floor to drop into the wye branches in the same stud bay below the floor, I then drop through the lower plates in a suitable location. Out of long-sweep 90s I come across the wall through bored studs horizontally and enter the branches of 3x2 or 4x2 combination wyes.

It's not often that you can get the smaller drains into the same lower stud bay vertically, then into wyes. Most of the time you'll be boring your way across studs. If you've got a 2x6 wall and the extra cost in materials for individual pipe runs to the building drain is within the job's budget, I suggest that you run individual fixture drains whenever you can. They work better that way.

BATHTUB-TO-VENT CONNECTIONS

VENT IN VALVE WALL

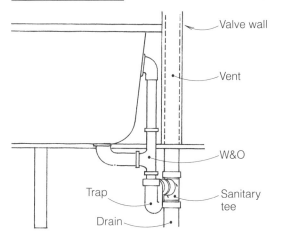

Valve wall

Vent

W&O

Trap

Sanitary tee

Drain

VENT IN SIDE WALL

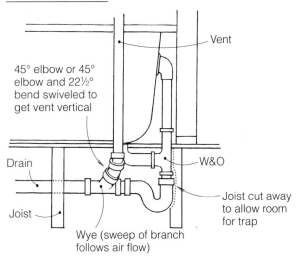

Vent

45° elbow or 45° elbow and 22½° bend swiveled to get vent vertical

Drain

W&O

Joist

Joist cut away to allow room for trap

Wye (sweep of branch follows air flow)

VENT DIRECTLY ABOVE DRAIN

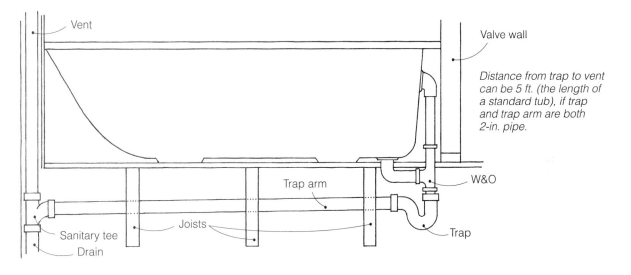

Vent

Valve wall

Distance from trap to vent can be 5 ft. (the length of a standard tub), if trap and trap arm are both 2-in. pipe.

Trap arm

W&O

Sanitary tee

Joists

Trap

Drain

PLUMBING A BATHTUB Installing a vent-to-drain tub pattern may be tricky because of the presence of the tub's waste and overflow (W&O) pipe, which will be right up against the lower plate of the valve wall, where the trap and trap arm might ordinarily go. So the patterns described on pp. 108-109 may need to be adapted.

In the most common arrangement, the vent and drain are in the valve wall (see the drawing at top left), and the vent connects to the drain through a sanitary tee. The trap angles backwards, and the trap arm is very short.

Sometimes the vent is installed in a side wall (see the drawing at top right). In this configuration the vent joins the trap at a wye rolled 45°. The joist may have to be gouged so the trap and the W&O will fit.

If the tub's valve wall is an interior, nonbearing wall not directly above a lower-floor wall, you may have to install the vent on another wall, as shown in the bottom drawing. If you are lucky, the floor joists will be running parallel to the long dimension of the tub so you can run the drain line between joists to the vent, without having to bore joists and use many couplings. But this is not always the case, and boring joists may be unavoidable. My major code says that a 2-in. dia. tub trap may be as far as 5 ft. away from its vent.

PLUMBING A SINK   At this stage, the only concern here is getting the proper height on the sanitary tee's branch (the trap and drain will be hooked up later, a process that is discussed in detail in my other book, *Installing and Repairing Plumbing Fixtures*). When the test-tee cleanout is installed below the sanitary tee on a cabinet-mounted sink drain, you cement (with ABS or PVC piping) a test tee onto the stub with the threaded branch about 8 in. off the finish floor (no-hub and copper have the same test-tee fitting). Then cut another piece of pipe, which when cemented into the upper inlet of the test tee, will leave the branch of the sanitary tee centered at about 17 in. off the floor.

If your community requires tubular-brass traps, you are better off using a sanitary tee that has a threaded branch. A threaded branch saves you some valuable inches that would be required to adapt from plastic to brass. You can thread a brass close nipple into the branch and then use a brass slip nut to grip a tubular brass trap arm.

If your area allows ABS or PVC plastic traps, you can use the standard solvent-weld tee and glue a short (6-in. to 12-in.) stub into the branch. After adding this branch stub, glue a test cap on its end. When I know I'm going to use ABS or PVC pipe for trap-arm stubs, I cut the stub and then glue the cap on one end and immediately place the cap end on the floor and let it set up in this vertical position before gluing it into the tee branch. This guarantees a good seal at the cap for the water test (see pp. 128-130). After the final water test you can cut off the test cap.

FIRST-FLOOR PLUMBING

With the upper-story iron drains brought down through drop walls and poked into the basement or crawl space, you can do the first-floor drain, waste and vent piping in ABS or PVC, using the vent-to-drain patterns discussed on pp. 100-109, if your local authority approves of plastic piping. As on the upper

story, start with with the toilet(s). Because wall thickness for the toilet drains is not a consideration here (except for a wall-hung toilet), you can use 4-in. pipe and fittings. You would measure the distance from the tank wall, provide proper side clearance and cut the closet flange hole on the first floor exactly as you did upstairs. You need to approximate the finish height as best you can with the flat surface of the closet flange. This might mean using shim material at this point to have a solid base to support the flange and pipe hanging below.

After the toilet drain and vent have been roughed in, continue with the other fixtures, just as you did on the upper story.

CONNECTING UPPER-STORY DRAINS TO THE BUILDING DRAIN

If you have chosen interior walls for drop walls (see the sidebar on p. 126), it is usually a snap to poke through into the crawl space or basement and join the drain to the building drain. The drawing on the facing page shows several options for making this vertical-to-horizontal connection. Most inspectors will require you to use a long-sweep 90 or at least a mid-sweep elbow at the bottom of a drop to bring the run horizontal.

If you joined one or all of the smaller drains to the toilet drain line inside the drop wall, then you might be bringing only one large-diameter drain line or one large and one small line through the lower plate or plates. If you brought all the lines down individually, they can be joined "in the dim." The height of the basement or crawl space, the ease of access and the surface condition of the grade will help you decide how many drains you want to be working with down there. If the conditions are particularly unpleasant, joining more of the drains together in the drop wall might be not such a bad idea.

If you are building your own house, you may have brought the tub/shower, lavatory basin, and bidet down to the crawl space or basement in individual 2-in. drains. But if you are plumbing for the typical penny-squeezing customer, you probably joined the smaller drains to the toilet drain in the walls or maybe under the second floor in a false ceiling. At any rate, you have any number of drains, of various diameters, poking beneath the lower plates of the

VERTICAL TO HORIZONTAL DRAIN CONNECTIONS

SIMPLE CHANGE OF DIRECTION

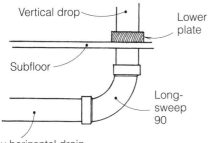

Vertical drop
Lower plate
Subfloor
Long-sweep 90
Any horizontal drain (including building drain)

BRANCH CONNECTIONS

#1

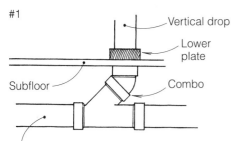

Vertical drop
Lower plate
Subfloor
Combo
Any horizontal drain (including building drain); could also be cleanout fitting

#2

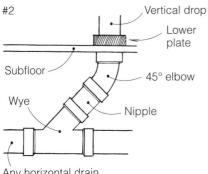

Vertical drop
Lower plate
Subfloor
45° elbow
Wye
Nipple
Any horizontal drain (including building drain)

#3

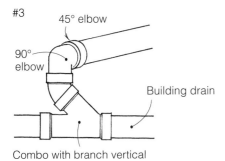

45° elbow
90° elbow
Building drain
Combo with branch vertical

#4

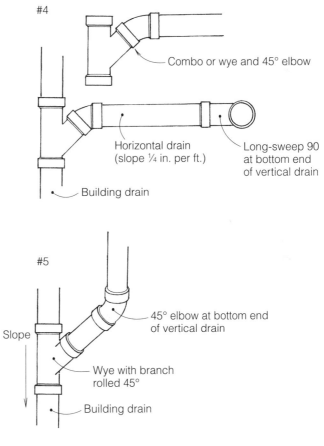

Combo or wye and 45° elbow
Horizontal drain (slope ¼ in. per ft.)
Long-sweep 90 at bottom end of vertical drain
Building drain

#5

Slope
45° elbow at bottom end of vertical drain
Wye with branch rolled 45°
Building drain

One key to a successful DWV system is selecting the drop walls, the walls through which the branch drains pass on their way to the building drain. When you are above the first floor and have the option of more than one wall for a drop wall, especially for a toilet drain, choose a wall where the noise will cause the least discomfort. Iron pipe is the quietest drainage piping. But when it is installed vertically inside a wall of a quiet room, like a dining room, you still might be able to detect some noise. I recommend that you use any wall except a dining-room wall for a drainage drop. But if a dining-room wall is the only possible location for a drain, insulating the pipe with rubber carpet padding will damp the noise somewhat.

Another thing to consider is whether you will have to bore through joists to get the branch drain over to the drop wall. Boring joists is tedious and time-consuming, and the more couplings you have to install, the greater the risk of leaks in the finished drain. The drawing below shows two possible drainage schemes for a bathtub. Alternative A calls for boring through several joists (and installing a coupling at each one) to reach the drop wall. Alternative B runs a branch drain parallel to the joists directly into the drop wall; no boring is required. Clearly, Alternative B is the better choice.

Remember that the thickness of the drop wall determines the diameter of the drain pipe you can use. A 2x6 or wider drop wall will accommodate 4-in. pipe; with a 2x4 drop wall you'll have to use 3-in. pipe and fittings.

BATHTUB DRAIN LAYOUT: WEIGHING ALTERNATIVES

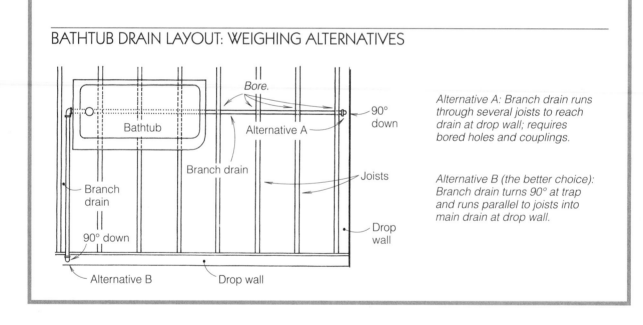

Alternative A: Branch drain runs through several joists to reach drain at drop wall; requires bored holes and couplings.

Alternative B (the better choice): Branch drain turns 90° at trap and runs parallel to joists into main drain at drop wall.

first floor. You will probably find that joining everything together can be done only on the horizontal or near horizontal, using one of the connections shown in the drawing on p. 125.

Begin by temporarily coupling a vertical long-sweep 90 to the bottom end of the second-story toilet drain and measure the available inches of fall from the center of the target municipal sewer outside of the building or the inlet stub of the septic system to the center of the fitting's lower leg. How far down you bring your second floor's vertical toilet drain below the first floor's subfloor before coupling the long-sweep 90 or other fittings depends upon how much height you have to work with (tight crawl space vs.

roomy basement) and how much usable space you want to have left after the piping is installed. Often the requirements of gravity and the limitations of the floor plans mean that pipes must run straight through the middle of otherwise valuable space. The spot to measure from may be a form sleeve through the foundation wall, a tunnel under the foundation wall's footing or a hole cut in the structure's exterior, above the foundation wall's mudsill. You need to see the exact placement of the exit hole from inside the building. From it, you can take measurements that will determine how close to the joists you must start the building drain to maintain acceptable fall.

If your toilet drain is only 3 in. in diameter, at the bottom of the drop you might choose to use a 3x4 no-hub coupling to join a vertical 4-in. long-sweep 90 to your 3-in. pipe drop if the horizontal run to the building drain is very long or if other drains have been joined to the toilet drain. You might want to join the 4-in. first-floor toilet drain into this horizontal 4-in. drain from the upstairs toilet by means of a wye on the flat, but this might be hard to do. It's more often the case that you will roll the wye branch to 45° up from vertical and then use ⅛ and ⅟₁₆ bends to connect to the branch line. If you have a very generous crawl space or basement, you can stack vertical combos, and out of long-sweep 90s on the individual drains, pull your branch drain lines into the combo branches.

CONNECTING THE DRAIN TO THE SEWER

Outside of the building wall, I join the building drain the the sewer with a two-way cleanout, or Kelly fitting, as shown in the drawing on p. 128. (Kelly fittings can sometimes also be installed just inside the building wall, with your local authority's approval.) A Kelly fitting should be installed horizontally, with the top branch as vertical as you can get it. Depending upon how deep the drain is in the ground, you might need to add an iron riser pipe to reach grade level — 3 in. to 4 in. above ground level is better. A no-hub blind plug, held in place by a no-hub coupling, is a common cap used to seal off the top cleanout entrance. Paint the plug white, so years from now you will be able to find it in the dark.

CONNECTIONS AT BRANCH DRAINS

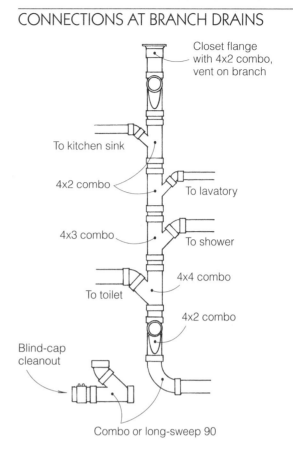

Closet flange with 4x2 combo, vent on branch

To kitchen sink

4x2 combo

To lavatory

4x3 combo

To shower

4x4 combo

To toilet

4x2 combo

Blind-cap cleanout

Combo or long-sweep 90

The building sewer starts at the outflow of the Kelly fitting and runs either to the municipal sewer or to your septic system. Depending upon where you live, the building sewer might be under a separate jurisdiction from the building department or whoever regulates the building process in your area. In many communities, sewers are inspected by someone other than the building inspector. If this is the case, I suggest that you consult with that person before doing any work; you might save yourself a lot of toil, trouble and money. The areas under the closest scrutiny will be your trench work (including bed preparation and depth) and the material for the sewer pipe and fittings.

Outside of the structure, many communities allow the use of terra-cotta pipe and fittings for building sewers, but for relatively short (30-ft. to 40-ft.) sewer runs, I'd advise you to use iron, ABS or PVC, which are much less susceptible to breakage. However, in

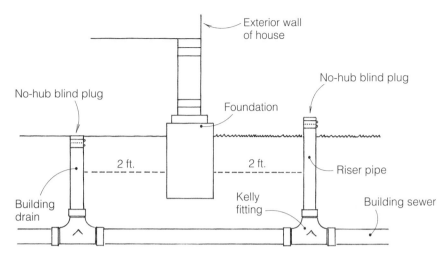

Locate Kelly fitting just inside or just outside foundation wall. With the approval of the local authority, a Kelly fitting may alo be located inside the foundation wall.

salty or otherwise corrosive soils terra-cotta is the best choice. On runs of hundreds of feet, the cost difference between iron and terra-cotta might tip the scales in favor of the latter. If you do use terra-cotta, lay it down on and then smother it in a generous bed of pea gravel or ½-in. gravel. Next to direct impact, earth movement is terra-cotta's biggest enemy; the gravel gives it a little breathing room, plus protection from digging tools.

TESTING THE DWV SYSTEM

When all the drains and vents are run, the system must be tested for leaks and approved by the local inspector before the walls can be closed in. Systems may be water tested or air tested. In water testing (the traditional test), all of the fixture connections are capped off, and water is introduced through a garden hose until it bubbles out the roof vent. Fittings are inspected for leaks, and leaky joints must be tightened or replaced. In air testing, all of the fixture connections and vents are capped off, and the system is filled with pressurized air. Leaks are identified by daubing the joints with soapy water and checking for bubbles, or by holding a candle flame near the joints and seeing if it remains steady.

Water tests are generally done on new houses in the stud and plywood stage; air testing is generally done on remodels. You might think that it would be easier to pass an air test than a water test, but I haven't found that to be the case. I prefer to use the water test because it's easier to locate leaks. The only time I do an air test is when I've done a remodel or an addition and a leak in the water test could prove disastrous to floors, ceilings or furnishings.

Whichever test you use, there are various types of caps for the fixture connections — these caps are available at your local plumbing supply. If you used a tapped tee for the kitchen sink and lavatory, a plastic or brass plug wrapped with Teflon tape is screwed into the female threads of the tee. For showers and tubs done in ABS or PVC, cemented test caps (see the top photo on the facing page) will suffice. Copper DWV is sealed with its version of the test cap. No-hub drains for tubs and shower that are still trapless can be sealed with no-hub blind plugs and no-hub couplings. If you have plumbed for a bidet, you will have either a unionless DWV trap in the floor to cap or a sanitary tee at the wall to plug.

A test cap cemented to open plastic pipes closes them off for testing.

For toilets, put a weenie into the closet bend or vertical drop and a test ball in front of that for extra protection, or try two test balls. A weenie (see the photo below) is an elongated inflatable testing device. There is one size of weenie for each size (diameter) of pipe. The weenie has a standard automotive tire/tube inflation valve, and you can pump it up with a tire pump. The test ball is a short version of the weenie.

Because there are a lot of duds among no-hub iron fittings compared to other materials, whenever I install a no-hub DWV system, I expect to replace a few fittings (see p. 91). This means that I'm always more apprehensive when it's testing time for this system.

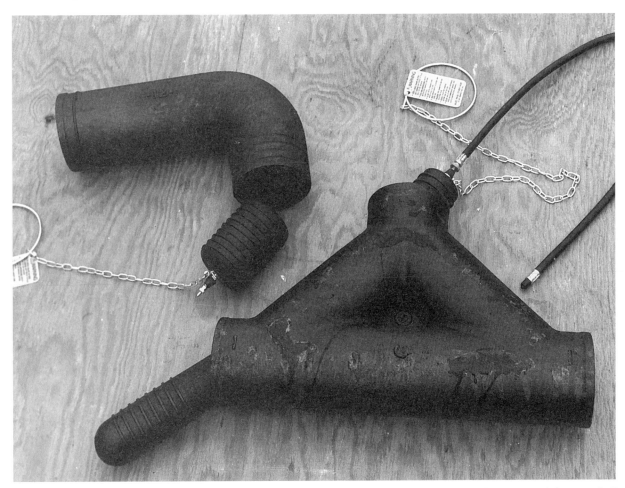

Weenies seal pipes for testing; the air can be introduced through an inflation valve by a bicycle pump. At left, a test ball plugs the end of a closet elbow; at right, a weenie seals a Kelly fitting.

A Jim cap with hose bibb lets you fill the DWV system with water when you test it for leaks.

On new construction, there is usually nothing but stud wall and plywood subfloor when I test, so a little water on the floor won't cause a lot of concern.

If the building sewer and inside-the-structure-plumbing will be inspected by the same person, I complete the sewer also if at all possible before calling for an inspection. If the sewer is under another jursidiction or being done by others or the trench for some reason can't be excavated until later, I cap off the building drain right outside the structure with a no-hub blind cap to which I have attached a hose bibb or use a commercially available Jim cap with a hose bibb (see the photo above). Onto this hose bibb I install a female hose to female hose adapter (see the drawing on p. 162). I then attach the male end of a garden hose and use someone's water (usually the neighbor's on a new house) to fill the system slowly, from the bottom up.

If you are testing the building's DWV system and the sewer line all at once, then you might have to get the hose up on the roof and into a vent stack if you have run a terra-cotta sewer. If your sewer is ABS or PVC, you can use the Jim cap with hose bibb at the lower end of the sewer line.

As the system fills with water, I pace around inside the building, tapping nervously on the vent pipes and upper-story drops. An empty no-hub pipe makes a clanky noise, one full of water makes a "dink-dink" noise. So I run around tapping, following the water up until I find a leak. Often it's only a coupling that needs a little snugging, and I can do it right then and leave the water on. If I encounter a big leak or a coupling that won't snug up tight enough to an out-of-round fitting, I have to shut the water off, drain just a little water out and replace the fitting. Most local authorities want to see the water bubble over the top edge of the vents on the roof, so I continue filling the system until water is streaming out the top of the lowest vent. Once I get to this stage (which might take hours), I leave the water inside the system for several hours more before letting it all out. Then, on the following work day, I call for my inspection and refill the system the morning of the day of the inspection.

To air test your DWV system, check with your local authority to see what the pressure requirement is. In my major and local code, I need to contain 5 psi or 10 in. of mercury for at least 15 minutes. To charge the system, you can use a standard, inexpensive 0-lb. to 35-lb. gas-test air gauge and valve. This apparatus has ¾-in. female iron pipe threads. You can use a nipple with a bell reducer to ¾ in., installed in a tapped sanitary tee for one of your lavatory basins or a kitchen sink. As for the water test, any ABS or PVC toilet lines can get a cemented test cap in the closet flange; you then need to install appropriate test caps for the piping material on your vent terminations and insert either inflatable test balls or weenies in iron toilet drains and the end of the building drain. When all the pipes are capped off or plugged, you can subject the system to 5 psi. Finding leaks at 5 psi requires more patience than at 15 to 30 psi, though. You can hold a candle flame close to each joint and check for movement in the flame, but it is nearly impossible to get all the way around all the joints in this fashion. You can also spray soapy water on the joints and check for bubbling.

DWV PLUMBING WITH A SLAB FOUNDATION

Plumbing with a slab foundation is trickier than plumbing with a crawl space or basement. Plumbing under a slab requires precise alignments and measurements, because pipe locations are virtually final once the slab is poured. For slab, all the the piping systems for the first floor have to be in place before the foundation is poured, as do the vents for toilets, tubs, showers and island sinks. The building drain and all the smaller drainage lines attached to it are stubbed off about 1 ft. above what will be finish slab height. Then, after the slab is poured, the above-slab plumbing can proceed as described earlier in this chapter. If you have stubbed off a pipe in the wrong location, however, it's unlikely that the floor plan will be changed to accommodate your error. You'll probably have to get a jackhammer, tear up the new floor and put the pipe where it belongs.

On slab foundations, finding the best wall for vents is a critical issue that must be decided before the pour. With off-the-ground construction, you can postpone your decision until after the walls are erected, but with slab, you cannot continue a pipe buried under concrete to another wall without busting up the concrete. (For the same reason, some codes require larger-diameter drainage lines on pipes that will be buried in concrete slabs, to reduce the likelihood of stoppages in these virtually inaccessible pipes.) When you review the plans, pay attention to any design changes affecting the walls that have been planned to be vent walls.

DWV CONNECTIONS UNDER SLAB

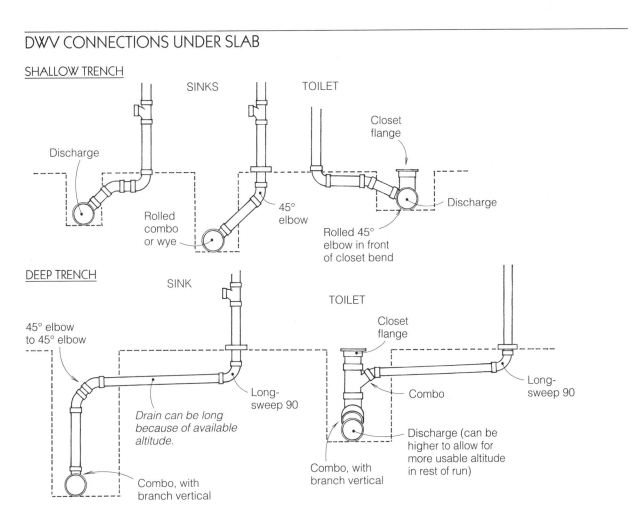

SHALLOW TRENCH

SINKS

TOILET

Discharge

Closet flange

Rolled combo or wye

45° elbow

Discharge

Rolled 45° elbow in front of closet bend

DEEP TRENCH

SINK

TOILET

45° elbow to 45° elbow

Closet flange

Combo

Long-sweep 90

Long-sweep 90

Drain can be long because of available altitude.

Combo, with branch vertical

Discharge (can be higher to allow for more usable altitude in rest of run)

Combo, with branch vertical

In cold climates, municipal sewers, building sewers and septic systems are located at greater depths than in mild climates. However, on slab floors in cold climates, the DWV pipes may or may not be deeper than in warm climates, because the house protects the piping directly below it from freezing. For this reason, some builders in cold climates keep their DWV up closer to the slab. If the piping is buried deeper, the plumber has the luxury of more altitude for the vent-to-drain patterns and drainage-pipe slope (see pp. 97-98).

On any slab foundation, the drainage piping goes in the soil, below the layer of aggregate and plastic-sheeting vapor barrier. On slabs where trenches are deep, getting all the piping safely in the soil is usually a simple matter, but with shallow trenches it can be a problem (some alternatives are shown in the drawing on p. 131). On slabs with shallow trenches, the primary plumbing concern is using vent-to-drain patterns with low altitude, especially for the farthest upstream toilet. You want that upper-end piping to afford as much altitude as possible for drain-line slope and additional downstream vent-to-drain patterns.

For slab, the choice of the correct toilet vent-to-drain pattern is critical (see pp. 100-105). The fifth and sixth patterns won't often work, except in cold climates where trenches can be deep, due to deeper burial of utility piping. For shallow-trench slab, that leaves four patterns, and of these, the first and second are the most useful. It is best to have as little piping as possible embedded in concrete. With the first, second and fourth patterns, most of the pipes can be kept out of the slab, with the exception of the vent. On bathrooms with interior toilet plumbing walls, the only embedded piping of any pattern is a 4-in. portion of vertical vent and a 4-in. section of pipe conecting whatever drainage fitting below slab hosts the closet flange above (for a discussion of closet flanges in slab foundations, see pp. 134-135).

PRELIMINARY LAYOUT

In installing DWV piping with slab foundations, you will need to be able to locate the walls before they are built. You will also have to install cleanouts, prepare for closet flanges, protect and support pipes that will pass through the concrete, and plan for tub boxes.

In order to establish the exact location of each fixture you will be plumbing, you measure out from the walls. However, when you are installing your below-slab DWV, no stud walls have been built yet, so you have to rely on string lines (see the drawing on the facing page), which must be laid out from the form boards set up for the pour. For interior walls, two string lines are used, one for each side of the wall, so there won't be any confusion as to which side of a string the wall is on. For exterior walls only one string, marking the inside of the wall, is necessary. I use masking tape on the strings so I can ink in the fixtures' centerlines.

Locate each fixture where the plan indicates, unless the planned position of a fixture is at odds with your major and local codes; in that case, nothing happens until the designer makes necessary changes. Using a framing square off my double strings to locate the spot, I drive a metal stake in the soil at the center of the toilet drain and at the center of the drain hole in the shower pan. Sink drains are measured off strings on walls perpendicular to the sink's plumbing wall. I use tape on the strings, centered on the sink drain locations and drive a stake there, too.

Tubs are a little more complex to lay out because the waste and overflow (W&O) that will be installed means that the drain hole in the tub will not be directly over the P-trap (see p. 109). Here you have to lay out a form, called a tub box, that will be set into the concrete to provide a void in the concrete for the trap. The drain discharge from the W&O is just about 1 in. in from the top edge of the valve end of the standard enameled-steel or cast-iron tub (tubs of other materials may differ somewhat). Measure from the actual fixture if you can, or work from a manufacturer's schematic as a second choice.

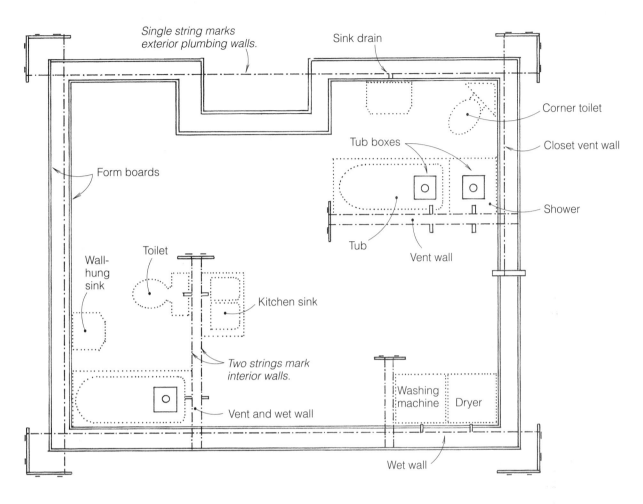

Single string marks exterior plumbing walls.

Sink drain

Corner toilet

Closet vent wall

Tub boxes

Shower

Form boards

Vent wall

Tub

Wall-hung sink

Toilet

Kitchen sink

Two strings mark interior walls.

Vent and wet wall

Washing machine

Dryer

Wet wall

The width of the tub is a critical factor in drain lay-out. Using the framing square, come off the strings that mark the inside wall of the tub at the valve end, half the tub width, and mark the valve-wall strings. For most standard tubs, this will be about 15 in. or 15½ in. Then measure out 12 in. with the framing square and drive a stake. This stake marks the center of the outside edge of the tub box, as measured from the valve wall. Both the shower and the tub get a tub box, and fractions of an inch are usually not an issue. You will read about staking tub boxes in position on p. 136.

CLEANOUTS

Installing an upper-terminus, below-slab cleanout on a shallow-trench slab can be difficult for a number of reasons. First, any cleanout extension pipe that you would need to install to connect the outside of the slab to the building drain will probably be considered a drainage line, so it would need the usual fall: ¼ in. per foot or maybe ⅛ in. per foot. You have to figure out how far it is to the edge of the slab from the up-permost toilet drain, and determine if the extension will be actually in the concrete by the time it gets to the outside. Another problem is the concrete itself:

CLEANOUTS ON A SLAB

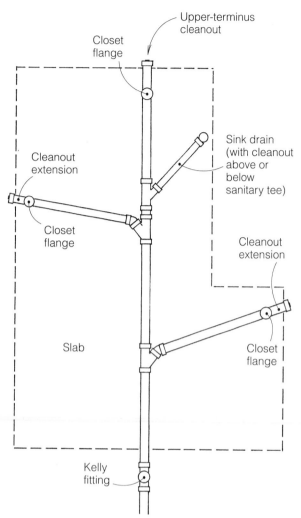

- Upper-terminus cleanout
- Closet flange
- Sink drain (with cleanout above or below sanitary tee)
- Cleanout extension
- Closet flange
- Cleanout extension
- Slab
- Closet flange
- Kelly fitting

A cardboard sleeve wrapped around a toilet riser keeps concrete far enough away to allow the closet flange to fit after the pour. (Photo by Bill Dane)

CLOSET-FLANGE PREPARATIONS

The safe way to plumb a toilet on slab is to leave the closet flange off and bring connecting pipe, called a riser, up past the finished floor height, and then trim down this pipe after the slab has been poured. When you bring the riser out of the cement, you need to provide a void around the pipe the thickness of the closet flange's hub, so that when you cut the riser down to the proper height, there will be room for the hub to be recessed into the slab. There are cardboard sleeves for this purpose (see the photo above), made by De Best Manufacturing.

In most parts of the country, below-slab DWV systems are run in ABS or PVC because they are less expensive to purchase and install than no-hub iron and DWV copper. Unlike iron and copper, the plastic pipes are resistant to damage from lime leaching from the concrete. I prefer to use PVC under the slab, because when it is properly joined with fresh primer and cement (see p. 61), the joint integrity is comparable to the welded joint of ABS, and PVC has more resistance to drain-cleaning chemicals, should they ever be used.

Regardless of the amount of rebar and mesh, concrete has a wondrous way of cracking directly over an embedded pipe.

Choosing a fitting to join the cleanout extension pipe to the main drain is also problematic. For maximum effectiveness, you want the cleanout to be upstream of any possible trouble spots in the line due to fitting use. Depending on the vent-to-drain pattern you have chosen, you may not be able to put the cleanout in the best place. This is why installing a Kelly fitting outside the edge of the slab saves lots of hassles at the upper end.

STAKING A CLOSET RISER

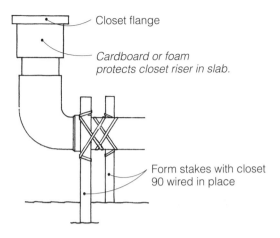

Closet flange

Cardboard or foam protects closet riser in slab.

Form stakes with closet 90 wired in place

You have two options for closet flanges with the ABS or PVC plastic piping. One type of flange has an adjustable metal rim, which you can rotate once it is glued in place to align the toilet with the plumbing wall. These flanges have holes in the metal rims for screwing to wooden floors; on slab I have used them successfuly by setting anchors into the slab. However, if you want to save a little time you should consider the second option, the solid plastic closet flange. This one-piece flange does not need to be anchored into the concrete, and can be installed either before or after the pour. When you install it after, make sure that you quickly position the closet-bolt slots at the proper position before the cement sets up. If you do not do this, then your toilet will be askew to the plumbing wall.

PIPE SUPPORT AND PROTECTION

The first-floor below-grade piping for a slab foundation has no framing members to hang pipes from; they must be supported or anchored in other ways. Plumbers installing piping for slab often drive pieces of rebar (reinforced steel) into the ground, to which they tie their piping with wire. (The rebar stakes stay in place during the pour and are not retrieved.) This takes time to do well but is very important. Other tradesmen often step on the piping during the pre-pour construction activity, and if it isn't securely positioned, the result may be negative fall on drain lines

or an incorrect finish height of the closet-flange riser. (If you have time to visit the site several hours before the pour begins, it is a good idea to give all of your piping a final check to make sure that it is still at the elevations you installed it.)

Rebar works, but I prefer to use form stakes (see the drawing at left) because they give much better support to the pipes. I find it difficult to wire the piping securely enough to rebar to resist at least some loss of altitude when someone stand or steps on the pipe. So at least around my toilet risers, I use form stakes, which I buy from the general contractor.

Your local authority will want you to wrap through-slab piping with a material that isolates the piping from the stresses of the slab and allows the pipe to expand and contract. Plumber's plastic foam is used for this purpose. You can also use cardboard, taped in place, if the local authority approves. Other materials that have been used in the past for this application are fiberglass building insulation and carpet padding, both natural fiber and foam rubber.

TUB BOXES

Tubs and shower stalls require a void in the slab around the stubbed-off and sealed drain line to provide room for installing the fixture's trap after the pour, when the fixtures are installed. If you live in a part of the country where slab is a common method of construction, the local plumbing suppliers usually sell the necessary forms and support products. One such product is the tub box, a rounded or square-sided plastic form with an open top, approximately 12 in. by 12 in. by 9 in. Tub boxes usually have one or more knockouts per side to poke the pipe through; they are sized for 1½-in. pipe (a tub-drain stub) and 2-in. pipe (a shower-stall drain). On the top edge of the box there are four tabs with holes, through which you can wire the box to rebar stakes to maintain its position. For each tub or shower, you purchase two individual forms, invert one on top of the other and wire them together (using the tabs) to prevent the form from being flooded with concrete. The plastic box has tapered sides so it can be removed from the slab after the pour, but most plumbers leave the lower form in the concrete.

TUB-BOX LAYOUT

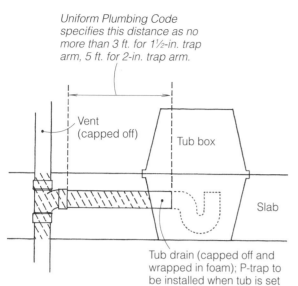

Uniform Plumbing Code specifies this distance as no more than 3 ft. for 1½-in. trap arm, 5 ft. for 2-in. trap arm.

Vent (capped off)

Tub box

Slab

Tub drain (capped off and wrapped in foam); P-trap to be installed when tub is set

For standard-sized cast-iron and enameled-steel tubs, the form is staked close to the centerline of the tub (allowing for swing room of the trap's J-bend) near the valve wall. For shower stalls, the location of the form depends on the location of the drain fitting in the floor, so you can't stake it until you know the type of shower being installed. For a manufactured unit you can check the drain-hole location on the schematic. For an on-site, custom-built pan, you had better find out where the tile setter plans to locate the drain. In either case, because you'll be working with the larger 2-in. trap, place the box so that the pan's drain hole is closer to one side, providing room to swing the J-bend of the trap to align under the drain opening.

If your shower stall is going to have a manufactured terrazzo or reinforced fiberglass floor (pan), try to find the pan if it is available, flip it over and look

A plastic pipe protected with white plastic foam wrapping (left) and a tub box secured to rebar (right) await the pour. (Photo by Bill Dane)

at the bottom. Most fiberglass stall units have floors that are off the finish slab level. And most fiberglass shower stalls require you to install a drain fitting in this pan. This drain fitting many times does protrude below the level of the finish slab. Many re-inforced fiberglass shower pans used in conjunction with other wall systems (tile, Corian, etc.) have an integral drain fitting that, in most cases, also protrudes below the level of the finish slab. The outside diameter of this "inverted mesa" can be upward of 8 in. If the pan is unavailable for reference, you might have to build your own, larger form.

TESTING THE BELOW-SLAB DWV

After the building drain is in place and the branched lines are stubbed off and sealed, I fill the system with water or air and test for leaks. As with off-the-ground plumbing (see pp. 128-130), ABS and PVC systems are easily sealed off with one-time-use plastic test caps, which are cemented in the pipe and or fittings. They can be broken out with a hammer when used with a fitting, or you can saw right behind them on a longer than required pipe stub when it is time to continue the piping. DWV copper uses a lightweight, one-time-use copper test cap, which is soldered over pipe stubs. It can be melted loose and discarded. No-hub iron uses an iron plug held in place with a no-hub band-seal coupling. All materials can be sealed off with varying degrees of success with air-expanded rubber test balls and weenies. These work best with smooth-wall pipe like the plastics and copper. With the orange-peel texture of no-hub pipe and fittings, you sometimes need more than one test ball or weenie.

With all but one of the stubs capped off, I attach a 10-ft. high section of test pipe to the only open stub and fill the system with water from a garden hose until it starts pouring out the top of the test pipe. Once the system holds tight on the water test, you can go ahead and install any fresh-water distribution piping that is being buried (see pp. 159-162).

After all the below-grade piping systems have been installed, tested, inspected and signed off by the local authority, I am usually sent on my way by the general contractor. I return weeks later when enough of the framing is up so that I can begin boring holes and stay relatively out of the way of other trades. From this point on, the work proceeds pretty much the same regardless of foundation type.

PLUMBING ABOVE THE SLAB

After the pour, all of your pipe stubs should be sticking up out of the plates about 6 in. If you haven't done it yet, you can go take your hammer and pop the test cap on the building drain and release the water out onto the ground or drain the water out through a hose. Once the water is out, cut the stubs below the caps. With your pocketknife, ream the rough edge on the inside and outside of the pipe. When wearing leather gloves, you can usually rub off the worst of the ridge left on the inside of the pipe's edge prior to reaming.

Now it is time to continue the DWV upward, beginning with sink drains up to and including the sanitary tee and any cleanout tees (also called test tees). If you have two stories to plumb, you should start with the toilet stubs and begin running drain lines up. With only a single story, you will need to continue up only to 1 ft. with smaller-diameter sink drains and vents. Connecting to these stubs is done most economically with cemented couplings if you are using ABS or PVC. A more expensive method is a same-material to same-material Mission coupling or other acceptable band seal for use in your area.

CHAPTER 6

WATER SUPPLY AND DISTRIBUTION LINES

This chapter covers roughing in the water service, supply and distribution system, which is usually done after the DWV system has been installed. This is also the time to do the rough plumbing for any solar or hot-water heating.

Compared to DWV plumbing, fresh-water supply plumbing is relatively straightforward. The goal is to extend branch lines up from the main supply line to each individual fixture that requires water. Once the pipe has been brought to the inside edge of the wall, it is stubbed off. When the entire system is installed and stubbed off, it can be pressure tested. For a slab foundation, the procedure is slightly different for piping that passes below, in and through the slab, as described on pp. 159-163.

For water-service piping (from the city main or well to the house), galvanized steel, copper or PVC may be used — check your local codes. For water-supply piping, copper is the material of choice. Poly-butylene is popular in some parts of the country, but not in others. Since copper is used everywhere, that is what I'll describe here. For information on PB joinery, see pp. 63-64 and carefully follow the manufacturer's instructions.

PLANNING AND LOGISTICS

In my area, fresh water moves through a service line from the city main to the meter, enters a building via a single pipe, the supply line, and is brought to the fixtures by branch lines. Most modest homes use a ¾-in. main service line from the water source to the building. With an average city-main pressure of 75 psi to 80 psi, the average two-story, two-and-a-half bath house, with laundry, is adequately served with a ¾-in. building service, a ¾-in. house supply and ¾-in. branch lines up to within a few feet of each fixture before reducing to ½-in. pipe (sizing supply pipes is dealt with more thoroughly on pp. 45-49). With a low city-main pressure, say, about 50 psi, or low well-pump pressure, the size of the service line and in-house main and branches would have to be increased.

The material you choose for your service line from the water source to the structure will determine how well you will need to protect the piping (galvanized steel is more resistant to impact than copper or plastic). Local codes dictate how deep in the ground this line should be buried to protect it from freezing. It is the plumber's responsibility to ensure that the path the piping takes is free of or shielded from obstructions — sharp rocks in the case of plastic or copper, and tree root systems and migrating soils in the case of all materials. In some jurisdictions the trench must be inspected by the local authority before any pipe is laid.

If the municipal meter is already in place or the well is already functioning, the plumber may or may not be required to get the local authority's permission to connect to the source. Most local codes require municipal meters to be protected by a meter box of an acceptable material and design. A conscientious plumber will make certain that the shutoff valve in front of (upstream from) the meter is easily accessible. Often it is a water-company employee and not the plumber who sets this box in place. If the shutoff valve is too close to the inside wall or partially covered by the box's wall, it may be impossible to operate. I have had to lift and reset many meter boxes installed by others so as to provide proper access to the shutoff valve.

INSULATION

Depending upon where you live, you could be facing the task of insulating hot-water lines or both hot- and cold-water lines. The insulation works best when it wraps all the way around the pipe with as few gaps as possible. When pipe is already held tight to a surface, you cannot get the insulation all the way around and under it. If you know beforehand that you will be insulating some piping, you can plan the path accordingly and choose an appropriate pipe-hanging method (see pp. 36-38).

There are a few approaches you can take to insulating supply pipes. On piping runs installed perpendicular to joist bottoms with two-hole straps, you can cut sections as long as the center-to-center distance of the joists and then do a little custom cutting on the ends of the insulation. You divide the edge width of the joist in two. On the end of the piece of insulation that has been cut to the center-to-center measurement, you use a sharp pocketknife or a sharp hacksaw blade and cut lengthwise at the top of the hole, half the edge thickness of the joist. Then cut down, vertically, to the end of the horizontal cut. When you peel and slip the piece onto the pipe, the inside edges should be tight to the joist and the long ends should cover half the pipe and strap. When second and additional pieces are joined end to end, the entire pipe will be covered. It is almost impossible to cut insulation without leaving a little rise or lump at the joints. I sometimes use a strip of plastic plumber's tape to straddle the joint and screw it down on each side of the now insulated pipe.

If you don't want to bother with all that custom notching, you can cut your lengths to fit tight between the joists and then cut little "rings" the width of the joist. You have to cut enough arc off each ring so it is not held off the pipe when installed in the gap. Cut a little at a time until the piece fits. I don't care for this method because it makes for extra joints and many times the piece falls out if it is not well glued with rubber cement. I'd rather deal with the screws and tape than with the smelly vapors and mess of the glue.

You can also hang long runs of pipe with plastic plumber's tape below the joists far enough that you know that you can later slide the insulation on and it will fit under the joist. Then by accurately mitering for branch lines (see the photo on p. 39), you can raise up the piping and cradle it almost snugly to the joists. With this approach you greatly reduce cutting time; there are fewer small gaps, and with the insulation against the wood structure, you greatly reduce pipe noise. This method takes a little more planning on pipe layout (discussed below), and because of the needs of framers and other tradespeople, you may have to do pressure tests on partial runs so that you can insulate and finish hanging the pipe at its proper height before moving on and getting out of someone else's way.

LAYING OUT PIPE RUNS

A major part of installing water-supply piping is boring a lot of holes. It's worth taking the time to lay them out accurately so you won't have to go back and fix them. I've got several chalklines for snapping the edges of wall studs when laying out the holes for supply pipes, but when the stud wall is open on both sides, I prefer to use a folding rule most of the time. Using the top of one of the hinges at a height that looks good and is above or below any outlets, junction boxes or switch boxes, I run down a run of studs, marking the edge. I've drilled a hole on the first fold-

ing leg; through this hole, with the rule folded at 90° I mark the approximate center of a 2x4 or 2x6 stud (see the photo above). It doesn't matter a lot if the mark is off dead center, as long as the holes align with each other, the same distance in from the edge of the wall. If one side of the wall has already been drywalled, I lay out the runs with a combination square.

One good reason to avoid long continuous runs with supply piping is that the spacing between studs limits the length of pipe that you can get through a hole. If the studs are a standard 16 in. on center, there is only 14½ in. of space between them, and you cannot get a piece much longer than that between two studs and started through a hole. This means having to use a large number of couplings on some runs.

One way to avoid long horizontal runs through studs is by bringing the water-supply line up through a stud bay and running branch lines through the attic, then dropping back down the individual fixture walls as required. If a second story, vaulted ceilings or a split-level floor plan prevent you from doing this, there's another approach to try (consult with the builder or foreman first). If one end of the plumbing wall happens to end at a hallway and there is no directly opposing in-line wall in the room across the hall, you can bore a hole right through the channel in the hallway and from the other room, send one or

two long sections through the bored holes. If your plumbing wall ends perpendicular to an exterior wall and the siding is still not up, you can bore right through the exterior wall and send one long pipe section through the bored hole run from outside.

The same strategy of boring in from outside of the room will work for second-story floor joists, although I always try to find another path for piping before boring joists. If you must bore joists, you should go through slightly above center. Holes for water pipes are rarely of sufficient diameter to have any impact on the strength of the bored joists. However, if holes have also been bored for a drain line, the joists' loading capacity may be affected and the builder may need to double one or more joists to compensate for the loss of strength.

If your fresh-water copper lavatory and toilet piping is going straight down from the upper plates, well away from either side of a stud, it's probably going to pass through at least one fire block. You can bore a hole in this block and thread your copper run through in two pieces and then couple them once they are vertical; alternatively, you can remove the block, bore it and then slide it over one full-length piece of copper and then nail the block back once the piping is in place.

NAIL PLATES

If your pipe runs in the stud wall within the reach of drywall nails or screws, you will need to protect it from punctures. The biggest enemy of copper pipe inside the structure is nails — especially drywall nails and the finish nails in wooden baseboard. To protect ½-in. and ¾-in. copper water lines, I install steel nail plates on the parts of the stud wall where the copper pipe passes through (see the photo below). Most nail plates have been punched and folded in such a way as to create little prongs that pound into the wood and keep the plate in place. The self-nailers used by electricians are thicker, harder steel and have stronger prongs than the nail plates sold in plumbing-supply houses, so you may want to buy your nail plates at an electrical-supply house.

Nail plates hammered into the edges of studs will protect copper pipe in the wall behind them from nail punctures.

WORKING AROUND OBSTACLES

Since the DWV piping is already in place when you install the water-supply pipes, you sometimes encounter obstacles in your path. For example, let's say you are running copper supply pipe horizontally through a 2x4 wall and need to cross a vertical 2-in. vent or drain in the stud wall. At this point you have several choices. If you had anticipated the problem before installing the vent or drain, you might have bored the holes for it as close as possible to one edge of the plates, then bored the holes for the water line as close as possible to the opposite edge. That way you can keep the water line running flat, but you have to protect it by putting a nail plate on each stud edge. On a 2x4 stud wall, you cannot go past a 2-in. vent or drain centered in the plate with ½-in. or ¾-in. pipe. So on slab, where the stub-up location cannot be changed, you might have to take the long way around, maybe going up to an attic space and then dropping back down to the fixture.

If you want better protection against nail damage, bore down the center of the stud wall, out of reach of drywall nails, until you come to the vent or drain, and then go around it. In water supply, it's preferable to use four 45° elbows (see the drawing on p. 142) and keep the 90° fittings to a minimum, because the tighter radius of the 90° fitting generates more friction in the system. As shown in the chart on p. 44, the equivalent length of a ½-in. dia. 90° elbow is 1 ft.; the equivalent length of a ½-in. dia. 45° elbow is .6 ft. However, you will soon discover that you cannot easily join opposing pipe runs with 45° elbows. It helps tremendously if you can shove backwards on one pipe run the distance of a fitting's socket in order to get the assembly in place. But in many cases when you have worked from two directions and are trying to join them, this cannot be done.

CIRCUMVENTING AN OBSTACLE

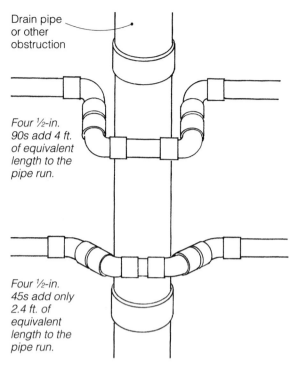

Drain pipe or other obstruction

Four ½-in. 90s add 4 ft. of equivalent length to the pipe run.

Four ½-in. 45s add only 2.4 ft. of equivalent length to the pipe run.

If your supply pipe has been sized to code minimums, the use of 90° fittings instead of 45s may mean inadequate water pressure or supply due to the additional friction in the pipe.

When space allows, a repair coupling (see p. 74) can be used between two sets of 45° fittings. But when going around an obstacle in a tight space, 90° fittings may be the simplest solution, and because I usually size the pipes more generously than the minimum code requirements, a few extra 90° fittings aren't going to make that much of a difference. Usually you can pry the ends of two opposing runs far enough off center to get a short piece of copper pipe (nipple) into the 90° elbow's opposing sockets.

For working in tight spots, the street fitting (it is made in both the 45° and 90° configuration) is especially helpful. As described on p. 17, one end of a street fitting is female, and the other has been reduced to the diameter of the male pipe, or streeted. That means you don't need the nipple. Because you can insert the streeted side of one fitting into the belled socket of another, you save the equivalent of one socket depth in the outside-to-outside distance of the two fittings. When fighting for fractions of an inch in order to stay inside the wall, street fittings can be life-savers.

DEALING WITH NOISY PIPES

In some systems, especially ones without a pressure-reducing valve (see p. 42), there are pounding noises in the piping as a fixture valve is shut off and the pipe bangs against the wall studs or joists. This annoying sound, called water hammer, is caused by water rushing against the shut valve. There are several ways to avoid the problem of water hammer, both homemade and commercial, and it makes sense to incorporate a solution into your plumbing system now, as you are installing it.

In some communities, it has been the practice to install a vertical section of pipe off one port of a tee, right in front of the angle stop or wall-hung faucet. This vertical pipe is then capped at the top end. The air chamber is intended to prevent water hammer in the piping by acting as a shock absorber. For best results the top end of the air chamber should be above the height of the valve being served, but if that isn't possible, the chamber will still have some effect. These air chambers are generally inside the walls for aesthetic reasons but that is a big mistake. Over time (maybe in as little as one year) these air chambers lose their captive air and fill up with water — once this happens the devices don't work. To make them functional again the building water supply must be shut off and the building drained down, evacuating the water in the chambers. Depending upon the path the fresh-water piping took to get to the fixture locations, you might have created a trap and it might not be possible to drain the water out of the chambers.

AIR-CHAMBER MANIFOLD

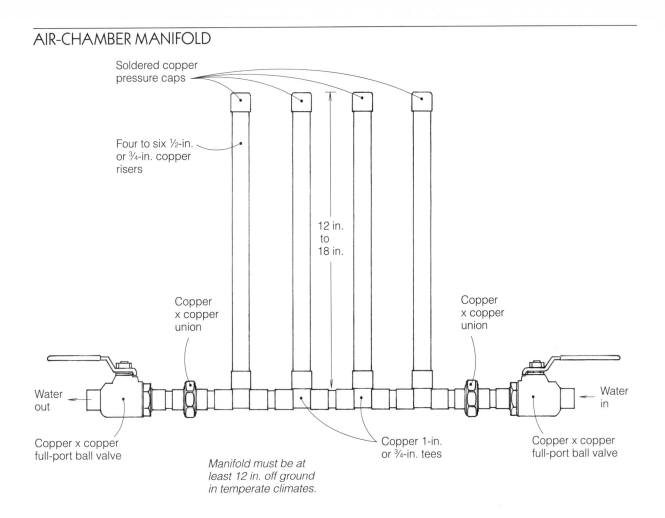

Soldered copper pressure caps

Four to six ½-in. or ¾-in. copper risers

12 in. to 18 in.

Copper x copper union

Copper x copper union

Water out

Copper x copper full-port ball valve

Manifold must be at least 12 in. off ground in temperate climates.

Copper 1-in. or ¾-in. tees

Water in

Copper x copper full-port ball valve

Air-chambered angle stops will usually always be accepted by the local authority. Another advantage is that the vent screw on these angle stops allows you to work at individual fixtures without shutting down the entire water supply to a building in order to drain it. For residential applications, one of the most best locations for an air-chambered angle stop is an automatic dishwasher. The air chamber saves the appliance's solenoid valve from a lot of hard use and extends the life of the appliance.

Instead of installing an air chamber at each and every fixture, it's much more practical to make up a manifold of as many as six air chambers on the main water line before or right after it enters the building,

as shown in the drawing above. In warm climates, if you locate the manifold outside the building, you can make the chambers as tall as you like and plant a shrub in front to camouflage them. On either side of the manifold, you solder a full-port ball valve (I've found it best to use valves made in the United States; economy off-shore brands are prone to leakage) and then inside each ball valve you solder a copper by copper union. This way, when the chambers fill with water and no longer stop the hammering noise, you merely shut off the ball valves, undo the unions and remove the manifold; then shake the water out, reinstall the manifold and turn the valves back on.

A commercial water-hammer arrestor is installed off a tee in the supply line, just behind the angle stop.

COMMERCIAL HAMMER ARRESTERS Commercial water-hammer arrestors are made by the Watts Co., Sioux Chief and others. These shock absorbers have male pipe threads on the bottom; some types have an air valve fitting on the top. You thread them into a tee branch (see the photo above) and they absorb line shock in the pipe. They have a diaphragm inside, and do not become waterlogged. A good location for this product is under a sink, where it could be threaded into an upturned branch of a brass tee, right in front of an angle stop.

INSTALLING THE WATER-SUPPLY PIPES

The plumber must run hot and cold branch lines from the main supply line to individual fixtures, including tubs and showers, lavatories, sinks, toilets, the water heater, the washing machine and exterior hose bibbs. If the fixtures have already been selected, look at the manufacturer's literature and schematics as you plan your plumbing runs, since not all makes and models have the same specifications. Installation of the individual fixtures is covered in detail in the companion volume to this book, *Installing and Repairing Plumbing Fixtures.*

The information presented in this section applies to off-the-ground construction; if you have a slab foundation, things are done a bit differently. Installing water-supply pipes with a slab foundation is discussed on pp. 159-163.

WATER-SUPPLY PIPING TO THE BATHROOM

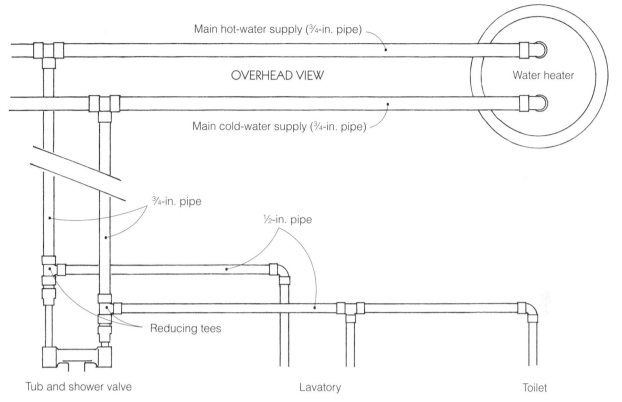

Main hot-water supply (¾-in. pipe)

OVERHEAD VIEW

Water heater

Main cold-water supply (¾-in. pipe)

¾-in. pipe

½-in. pipe

Reducing tees

Tub and shower valve

Lavatory

Toilet

TUB AND SHOWER

Everyone wants plentiful and controllable hot and cold water in the bathtub and shower, so let's begin the discussion of water supply with a discussion of these fixtures. The tub or tub and shower valve is the target for the main hot and cold water-service branches to the bathroom. If the framing and the absence of other trade hardware (ducts and wiring) allow, I maintain the full-size ¾-in. pipe run up to within a short distance of the tub and shower valve, where I switch to ½-in. pipe for the lines to the lavatory, toilet and bidet (if there is one). When reducing pipe size at the tub/shower valve, I often use a reducing tee. This copper by copper fitting has sockets that slip over pipe.

Try to find out the make and model of the tub and shower valve that will be used, so you can do the rough plumbing accordingly. You also need to know what material and what thickness the finish-wall system will be so you can secure the valve at the proper depth inside the wall.

If you are lucky, the valve wall will have been framed with off-center studs, leaving a center bay where blocking can be added for securing the tub and shower valve. If not, you have to chop out the central stud to get the valve in place, and then there is no good way to anchor the valve because there's nothing to attach the blocking to. You have to make do with plumber's tape.

There are many sizes and styles of tub and shower valves, and they cannot all be mounted the same way. Some brands are more demanding than others in their installation depth. Usually nothing can be done for a valve that has been roughed in too far out of the wall, but that situation is rare. The more common problem is installing the valve too deep because of a misunderstanding about the finish-wall material or a change in plan for wall finish after the valve has been roughed in. A few valve manufacturers offer stem extensions, which allow valves that are installed too deep in the wall to be used without destroying the finish wall and moving the valve forward.

There is no standard height off the floor at which you set and anchor the valve inside the wall. The height of the valve in the wall is usually best arrived at by having the intended users stand at the tub's or shower's valve wall and let their arms hang comfortably at their sides, then reach for an imaginary valve handle. I find 32 in. to 34 in. off the floor a comfortable height. If the users rarely shower, have them sit on the floor and reach forward, as if to add more hot water to the bath. Mark the wall at this height and then add another inch or two to account for height of the tub's floor off the subfloor.

The shower-arm fitting (a ½-in. copper by FIP drop-eared 90) is usually installed inside the wall at a height between 65 in. and 78 in., but you don't have to put it there. You should consider the height of the users, especially if they are exceptionally short or tall.

Most rough-in books produced by major fixture manufacturers (free to those who ask) recommend that the tub's filler spout be set about 4 in. above the top edge of the fixture, but you can put it at whatever height seems best. Be sure that you use a drop-eared 90 to anchor the spout. The tub spout is often grabbed by folks who have lost their footing or who are afraid of doing so. You want a tub spout that is anchored in the wall like the rock of Gibraltar.

The need for a strong installation automatically precludes a plastic fitting, as well as the use of copper tubing as a transitional pipe material. The best spouts are made from solid brass and thread onto brass pipe nipples. Do not use spouts (almost always made of plastic or die-cast metal) that slip over ½-in. copper tube and are held in place with an Allen screw. If someone grabbed a lightweight spout mounted on copper pipe to stop from falling, the spout might break off at the wall, pulling the tube out with it, and injury might result.

INSTALLING A TUB AND SHOWER VALVE   As shown in the drawing on the facing page, framing for the tub plumbing requires the installation of three or four pieces of blocking: one at the spout, one at and possibly above the valve, and one at the shower head. I begin with the one at the valve. How far back in the wall this block is installed determines whether the valve trim (handles and escutcheons) will fit and

function properly. You need to know the distance from the face of the studs to the finish-wall surface to position this block properly.

The block at the valve may be installed flush with the front edge of the studs or farther back in the stud bay, depending on the size and shape of the valve, the depth of the wall (2x4 or 2x6) and the positions of the water inlets and outlets. If there is plenty of room depthwise in the stud bay, I often install the valve to the face of a flat block (preferably a 2x6 using pure copper, two-hole plumbing tube straps. (In this situation I can usually use these straps around the tub and shower riser ports too.) With the valve mounted this way (Alternative A in the drawing), the rest of the installation gets only two additional blocks: one for the tub spout's drop-eared 90 and one for the shower arm's drop-eared 90.

If there isn't enough room to attach the valve to the face of the blocking, I often bore holes for the ½-in. supply pipes vertically through the block. The block can be flat (Alternative B) or on edge (Alternative C). When the block is flat, the long holes add rigidity to the installation. When the block is on edge, I locate the holes closer to the back edge of the block, so that when the block is installed, the holes are as far as possible from the face of the stud (this maneuver can gain you an inch over fastening the pipes with drop-eared 90s). However, with alternatives B and C, if you are installing both a tub and shower valve and a shower arm, the shower riser will be unsupported and the setup will require an additional block above the valve, just above the shower riser port. Whether you bore this block too or mount a drop-eared 90 to the face, flat side of the block at the shower head and use 45° fittings and offset the discharge piping to make the drop-eared 90 plumb with the riser port on the valve makes little difference. Take the easiest path.

If I have installed only one block below the valve (Alternative A), I then do the tub spout drop-eared 90's fitting block and then the shower-arm drop-eared 90's block. Whether the blocks are installed first and then the valve or whether you need to attach the valve and then install the block/valve as one unit depends on the framing of the individual job. Sometimes prefabricating the valve/block assembly helps get the fasteners in the best possible locations, something that isn't possible when you have to make the valve mounting fit pre-installed blocks.

SUPPLY PIPING AT HEAD OF TUB: THREE ALTERNATIVES

A. THREE BLOCKS (ALL FLAT)

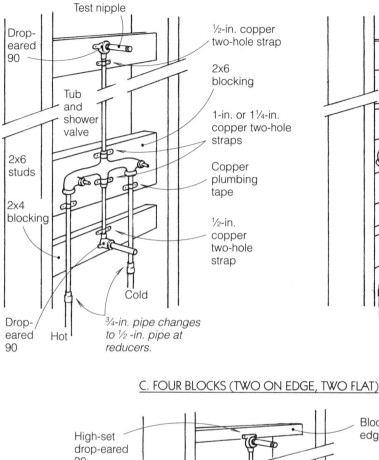

Test nipple

Drop-eared 90

½-in. copper two-hole strap

2x6 blocking

Tub and shower valve

1-in. or 1¼-in. copper two-hole straps

2x6 studs

Copper plumbing tape

2x4 blocking

½-in. copper two-hole strap

Drop-eared 90

Hot

Cold

¾-in. pipe changes to ½-in. pipe at reducers.

B. FOUR BLOCKS (ALL FLAT)

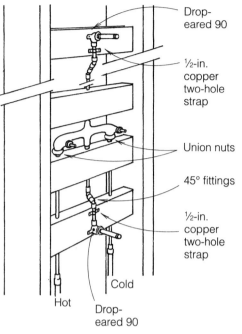

Drop-eared 90

½-in. copper two-hole strap

Union nuts

45° fittings

½-in. copper two-hole strap

Hot

Cold

Drop-eared 90

C. FOUR BLOCKS (TWO ON EDGE, TWO FLAT)

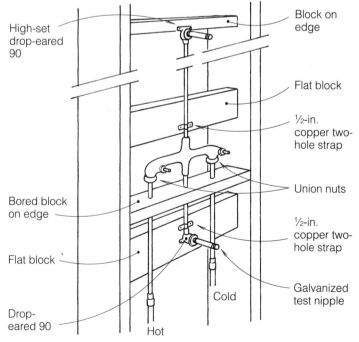

High-set drop-eared 90

Block on edge

Flat block

½-in. copper two-hole strap

Union nuts

Bored block on edge

½-in. copper two-hole strap

Flat block

Galvanized test nipple

Drop-eared 90

Hot

Cold

The ½-in. copper by FIP brass drop-eared 90 that I bring up above and below the valve for the shower-arm and tub-spout nipples has a hole in each ear, through which I then screw a 1½-in. long No. 6 or No. 8 stainless-steel sheet-metal screw into the block. Then I use additional pipe straps on the copper pipe, screwing them into the blocking. Into the drop-eared fitting I thread a galvanized nipple and cap for water-test purposes (see p. 158). When I return weeks to months later to set the fixture and install the trim, I replace the galvanized nipples with brass ones of the appropriate length. That brings the back of the filler spout up against the finish wall with as many threads meshing as possible in the back of the spout and drop-eared fitting, for maximum strength.

You can purchase tub/shower and shower valves with or without unions at the hot and cold inlet ports. Most standard two- and three-handle, 8-in. spread valves have unions that accept threaded pipe. The ports for the shower riser and filler spout drop will also be tapped for ½-in. iron pipe. Expensive valves might also include two separate union halves made for soldering ½-in. copper tube into them. On valves without unions, all connections are usually female copper, into which you directly solder your copper pipe.

On a 2x4 tub or shower wall, the valve is usually so big that you can't use flat 2x4 or 2x6 blocking to anchor it and still be the proper distance inside the wall. For support blocking here, you could add a piece of 1x6 or 1x8, preferably full rough thickness and screwed in place. For a two-handle valve, three or four 1-in. copper pipe straps usually work well to screw it to the blocking. For a single-handle mixing valve, which lacks the usual good attachment spots of a two- or three-handled valve, I sometimes use copper plumber's tape to cross-brace it across the front and screw to the blocking.

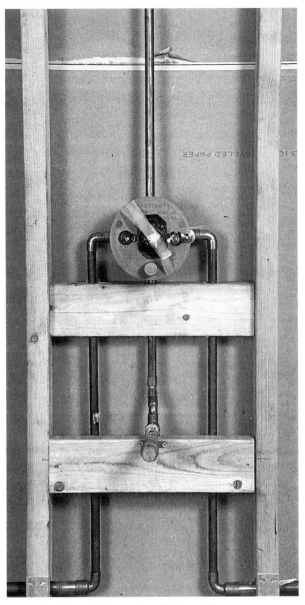

A roughed-in tub/shower installation with a single-handle valve.

SINGLE-HANDLE VS. TWO-HANDLE VALVES Single-handle mixing valves (see the photo above right) can be purchased with FIP threaded ports for use with threaded iron pipe or with smooth-bore ports for sweating onto ½-in. copper tube. You can always use a valve with the threaded ports for copper by first screwing copper by MIP adapters into the valve and then soldering copper tubing into the smooth side of

the adapters. When I do this, I first cut two pieces of tube, longer than they have to be, and solder on the adapters. I wrap Teflon tape on the MIPs' threads and put a little pipe dope in the valve's threads and screw them in. I hold the valve up in the wall and mark the two pipes where they need to be trimmed. The valve, which is a big hunk of brass, absorbs a lot of heat before the adapters get hot enough to melt the solder.

Soldering this way takes time; a faster way is first to solder adapters to the tube and then assemble. This second method has the advantage of subjecting the valve to as little heat as possible. Many first-timers overheat the valve when soldering to it, thereby destroying its internal plastic cartridges or rubber seals or both.

When you purchase a valve for shower only, the manufacturer often gives you a tub/shower valve and throws in a brass plug for you to add to the lower filler port, which will not be used. In this case, I go ahead and add an MIP adapter and solder a stub into it and then cap the stub. I then add another flat block below the valve and use several copper pipe straps, screwed over this added, capped stub, to secure the valve rigidly in place.

Two-handle valves often come with some sort of sleeves to protect protruding stem assemblies until the valve trim can be installed. These sleeves are shoved in place, but don't trust them to stay there. The tilesetters won't care if the sleeves come off; they'll just goop up the valve and spoil the finish and ruin the trim threads. Securely tape the sleeves in place, now (see the photo at right). Single-handle valves come with a plaster guard installed to the valve with one screw. The only vulnerable trim on this valve is the sleeve around the handle stem. Some valves have an additional plastic or cardboard dome covering the stem, held in place with a screw. Others might have a small cardboard cap just shoved over the stem. Tape whichever type you have to the plastic plaster guard.

If the valve hasn't yet been chosen yet, which is often the case, it's best to do as little as possible until you know what make and model you'll be dealing with. Single-handle valves are much smaller than two-handle valves. The supply connections (unions) for two-handle valves are always vertical, and some models and brands allow you to feed from below or above, usually after moving a few parts, as noted in the manufacturer's instructions. Single-handle valves usually feed from horizontal ports, but some antiscald models, because of internal controls, might have unusually placed ports.

If you aren't sure which valve you are going to use, just stub off your water, coming up from the lower plates or coming down from the upper plates, 8 in.

A roughed-in tub/shower installation with a two-handle valve. Taping the sleeves in place protects them from damage when the wall is tiled.

on center and several inches past whatever finish height you have chosen. If you use a two-handle valve, you can go straight into the unions. If you use a simple single-handle valve, you can add a 90° elbow off the stubs and go straight into the ports. If you end up with a temperature-compensating (antiscald) single-handle valve, you'll have enough room to reach the unorthodox locations if you keep the supply tubes on 8-in. centers.

LAVATORY AND TOILET

For the supply piping in the rest of the bathroom, I run reduced-diameter (½-in.) lines from the main (¾-in.) branches to the lavatory and toilet. To reduce pipe sizes, I might use ¾x¾x½ reducing tees or straight ¾-in. tees with a ¾ by fitting by ½ copper reducer at the branch. (This reducer is streeted, so no nipple is needed to join it to the branch of the tee.) You insert the ¾-in. portion right into the branch of the ¾-in. tee and then come out of the ½-in. side with ½-in. pipe. Typical rough-in piping for the lavatory and toilet is shown in the drawing below.

The targets for the toilet and lavatory branch lines are copper by FIP drop-eared 90s, and you have to install blocking to secure these to. Usually there are enough short pieces of framing lumber on the job's scrap pile for this purpose. Angle stops that thread into the drop-eared 90s (see *Installing and Repairing Plumbing Fixtures,* the companion volume to this book) regulate the water supply to the sink faucets

and the toilet fill valve, and almost all local codes require that they be installed as a means for shutting off water to these fixtures.

If you are moving horizontally in a wall, boring studs to reach your location, once at the fixture location, the branch of your tees will either point down or up depending whether you are above or below the angle-stop height. By using 45° elbows or street 45° elbows, you can move your pipe forward or backward in the wall so that the drop or riser reaches the blocking for the angle stops in the same plane. Now solder ½-in. copper by FIP brass drop-eared or high-set 90s to the angle-stop stubs. (I prefer the drop-eared fitting to the high-set fitting because the former is stronger.) Then screw the 90s to the block.

For a lavatory, an 8-in. spread is the standard distance between the hot and cold water supplies. Many times structural obstacles make this spread difficult to achieve. If you are plumbing for a cabinet-mounted lavatory, you need to know now if conventionally

SUPPLY PIPING FOR LAVATORY AND TOILET

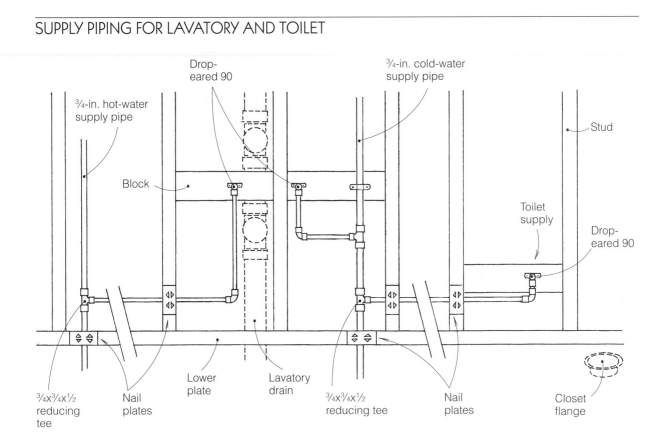

³⁄₄-in. hot-water supply pipe

Drop-eared 90

³⁄₄-in. cold-water supply pipe

Stud

Block

Toilet supply

Drop-eared 90

¾x¾x½ reducing tee

Nail plates

Lower plate

Lavatory drain

¾x¾x½ reducing tee

Nail plates

Closet flange

placed angle stops will interfere with any drawers in the cabinet. You might have to set the eared 90s at an unconventional height and spread. Often no one has given this sort of detail any thought, and you or some finish carpenter might end up shortening drawers or chopping out some integral backing support to get the cabinet set. If your major supply runs are crossing in an attic space and your tees are not at perfect right angles to fixture locations, you can point the branch of your tee vertical, then solder in a street 90° elbow or short nipple and a copper by copper 90° elbow and swing fit, coming out of the top, horizontal socket of the fitting in a direction that will let you drop down through the upper plate to the blocking.

PEDESTAL SINK Pedestal sinks pose a problem because the supply pipes and angle stops will be visible, and if the stops are not symmetrical with the center of the pedestal, the installation will look second rate. If you plan to install a pedestal sink, find out the make and model. Most pedestal sinks made by reputable manufacturers have a rough-in schematic published in the company's plumbing-fixture rough-in measurements book. If your wholesale plumbing supplier is reluctant to give you a free copy of this book, ask for a photocopy of the schematic in question. Most pedestal sinks have recommended water-supply stub spreads at the standard 8-in. distance apart and an average height from finish floor of about 20 in., but the stub spread does vary with the styling.

KITCHEN SINK

When running water lines to the kitchen sink (see the drawing on p. 152), anchor the ½-in. copper by FIP drop-eared 90s to studs, if possible. A good height for the kitchen-sink angle stops is 16 in. to 19 in. off the floor. There is a ½-in. copper by FIP high-set winged brass 90° elbow that has ears fastened to the top face of the FIP opening. Where it is not feasible to run cross blocking between studs because of vent or drain lines, and you have studs in appropriate locations where you want your angle stops, you can use the high-set winged 90° elbows horizontally, screwed to the face of the studs, as shown in Detail A of the drawing. Another alternative (Detail B) is to bore

into the center edge of the stud about 1 in., using a 2-in. or larger Milwaukee or Lenox boring bit in a right-angle drill, and then screw the wings of a drop-eared 90 on the centerline of the stud; on that fitting goes a street 90 elbow or short stub of copper and a copper by copper 90° fitting to turn the line either up or down, whichever direction you need for approaching the angle-stop location. Boring into the stud recesses the fitting back far enough so that the pipe is in somewhat from the edge of the lower upper plate, but nail plates (see p. 141) are still a good idea.

Whatever fitting you use, you eventually end up with a brass tee and three angle stops on the supply lines under the sink. On the cold supply, one of these angle stops will serve the kitchen-sink faucet and the other (a ½-in. FIP by ¼-in. compression fitting) will serve the refrigerator's ice maker.

HOT-WATER DISPENSER You can save yourself a lot of headaches later if you do the rough plumbing now for a hot-water dispenser, whether or not the plans call for one. (This option is not shown in the drawing on p. 152.) While you are installing the kitchen water-supply pipes, you can tee off the cold supply and provide another copper by FIP eared 90 and test nipple for the angle stop for a future instant hot-water appliance. Add the other winged 90 (and nipple) at some point left of center of the proposed dispenser on the wall where it will be far enough off the floor as not to interfere with a possible drawer.

There are two basic designs of instant hot-water dispensers. One (which I prefer) has a one-piece spout and tank that hangs from a 1¼-in. hole through the sink or countertop, with only one supply tube. The other type has a separate spout and tank, and the plumber has to deal with a cold-supply tube, a hot-supply tube between the tank and the spout, and a vent line. Whichever is chosen, the spout will need to occupy one of the predrilled holes in a cast-iron or porcelain sink deck. If the sink you are installing is stainless steel, you can punch an additional hole for the dispenser.

SUPPLY PIPING FOR KITCHEN SINK

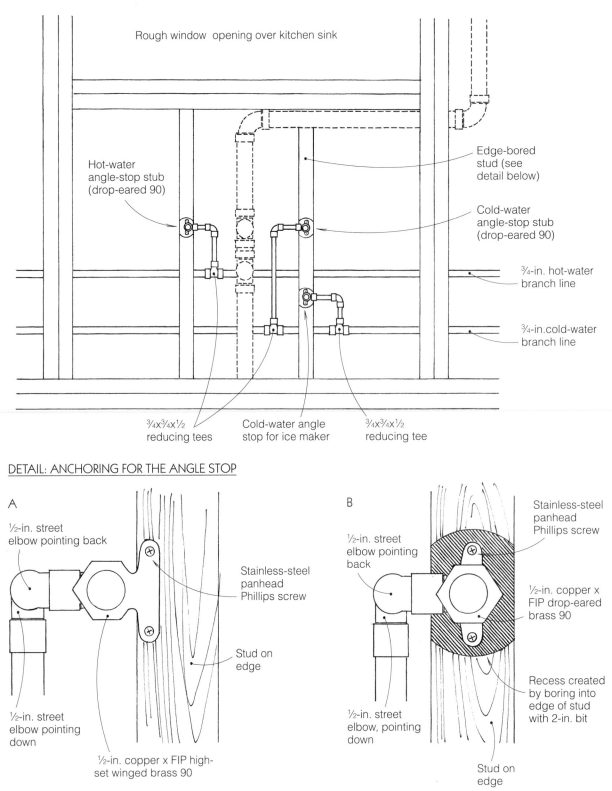

Rough window opening over kitchen sink

Hot-water angle-stop stub (drop-eared 90)

Edge-bored stud (see detail below)

Cold-water angle-stop stub (drop-eared 90)

¾-in. hot-water branch line

¾-in. cold-water branch line

¾x¾x½ reducing tees

Cold-water angle stop for ice maker

¾x¾x½ reducing tee

DETAIL: ANCHORING FOR THE ANGLE STOP

A

½-in. street elbow pointing back

Stainless-steel panhead Phillips screw

Stud on edge

½-in. street elbow pointing down

½-in. copper x FIP high-set winged brass 90

B

½-in. street elbow pointing back

Stainless-steel panhead Phillips screw

½-in. copper x FIP drop-eared brass 90

Recess created by boring into edge of stud with 2-in. bit

½-in. street elbow, pointing down

Stud on edge

WATER HEATER

Rough plumbing for a water heater depends on the type of appliance. All types — gas, electric and oil fired — require a cold-water branch line from the main supply and a hot-water outlet from the tank, and gas-fired heaters also require supply and vent piping for the gas as well, which I plumb at this time. (Fuel-gas supply piping is discussed in Chapter 7; venting a gas-fired water heater is discussed in Chapter 8). Oil-fired water heaters also require supply and vent piping for the oil. The only difference between oil- and gas-fired water heaters is that the former require triple-wall chimney vent.

The placement and combustion-air requirements of gas-burning water heaters are governed by code. My major code does not allow the placement of fuel-burning water heaters in any room designed for or used for sleeping, any bathroom or closet or confined space opening into any bathroom or bedroom. Local codes often prohibit the placement of a fuel-burning water heater under a stairway. But there are no restrictions about placing a fuel-burning water heater in a kitchen. In days gone by many were placed there, near the gas range to take advantage of its gas line and vent. Most new construction today does not waste the space in the kitchen that the heater would occupy, and it is placed elsewhere.

My major code further stipulates that if you place the fuel-burning water heater in an enclosure, you must provide it with combustion air. Without sufficient combusion air the burner flame will produce soot, which can build up and clog the flue inside the water heater. When this happens the burner flame seeks out a new path, usually out the bottom and up the outside of the heater — a very dangerous situation. My major code offers guidelines for ensuring an adequate amount of combustion air. If the square area of the enclosure is less than twice the the square area of the bottom of the water heater, 2 sq. in. of duct or opening must be provided for each 1,000 input Btus for a water heater that consumes up to 500,000 Btus. (The average 50-gal. residential water heater consumes 50,000 Btus.) When the enclosure has a floor area more than twice the area of the bottom of the water heater, 1 sq. in. of duct or opening must be provided for each 1,000 input Btus.

If your structure was designed by a reputable local architect, engineer or designer, code requirements for the water heater have probably been met. If you are the designer of the structure and you are not fully familiar with your local codes, then you should check with your local authority as to the acceptability of the chosen location. Most 20-gal. to 50-gal. residential holding-tank water heaters are close to 5 ft. high, and I like to leave a little space, at least 1 ft., between the water stubs and the top of the heater. If your water heater is located in a closet inside the house, you'd be safe in stubbing the ¾-in. copper lines at about 6 ft. from the finished floor. If the water heater is located in the garage, most codes want it to be at least 18 in. off the floor on a stand or platform, or even boxed in with an access door to the outside only.

When selecting a location for the water heater, remember that someday it will probably leak — maybe out the bottom of a rusted-out tank, from around the threaded connections on the top or from the drain cock. Whatever the source of the leak, the water will end up on the floor under the heater. Try to locate the water heater where a leak would do no damage. You can also consider placing the heater in a water-heater pan (called a Smitty pan in my area). This is a round aluminum pan about 2½ in. deep with a PVC plastic drain fitting in its wall. Into this fitting you can cement a ¾-in. plastic drain line to carry off the water to an appropriate location (outside at grade).

When rough plumbing the building's water supply, you stub out for the water heater, which will be installed later. I stub my cold-water intake and hot-water outlet lines about 12 in. apart and sometimes

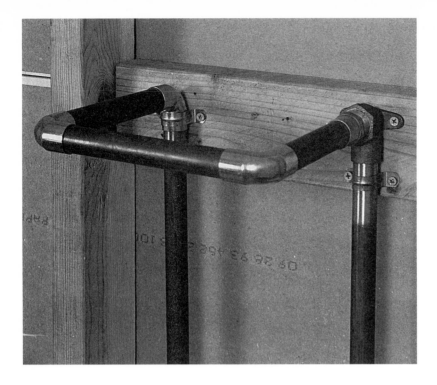

In this mockup, water-heater lines have been temporarily linked with 90° elbows and a length of connecting pipe for testing.

then use two 90° elbows and a piece of pipe between them to connect the two together about 12 in. from the wall, as shown in the photo above. (This is one way of providing the necessary link to get cold test water into the hot-water piping even though there is no actual heater in place yet.) Sometimes I just use a water flex to do the same thing. You can also cap off each pipe individually and then open both the hot and cold sides of an installed tub/shower valve to allow the cold water to enter the hot system for test purposes. If you have no valve stubbed in yet, you could connect the hot and cold sink supplies, as described on pp. 162-163 for testing the below-slab portion of the system.

LOOP LINES   If you are plumbing a new home (all the walls are now open stud) and there are some far-flung bathrooms, you might consider running an additional hot-water line from the water-heater location to the most distant hot-water demand. On this line, called a loop line, you can add a circulating pump, which will keep the waiting time for hot water to a minimum. When weighed against years of aggravation and water wasted, the extra cost may be a bargain. The loop line need not be any larger than ½-in.

pipe (⅝ in. O.D.). The far end of the loop plugs into a tee on the main hot-water service branch, as close as possible to the most important hot-water fixture, usually the tub/shower. The other end of the loop plugs into the suction side of the circulating pump, and the delivery side connects to the water heater.

To hook up the loop system (see the drawing on the facing page), you remove (unthread) the drain cock that comes on new heaters and install a ¾x3 brass nipple in the threaded opening. On this nipple you thread a ¾-in. brass tee. Into the opposing barrel of the tee, thread a ¾-in. brass male drain cock; into the branch of the tee you thread another short brass nipple, and then thread a full-port ball valve onto this nipple. Out the other side of the valve, you add another brass nipple, then a brass union and from there, solder up your piping to the delivery port of the pump. On the suction side of the pump, you have another union before the final ball valve. The loop is plugged into one side of this valve. These valves allow you to shut off the flow in the loop and remove the circulating pump for service, without having to drain down the water heater and house hot-water lines.

LOOP LINE FROM WATER HEATER TO TUB/SHOWER

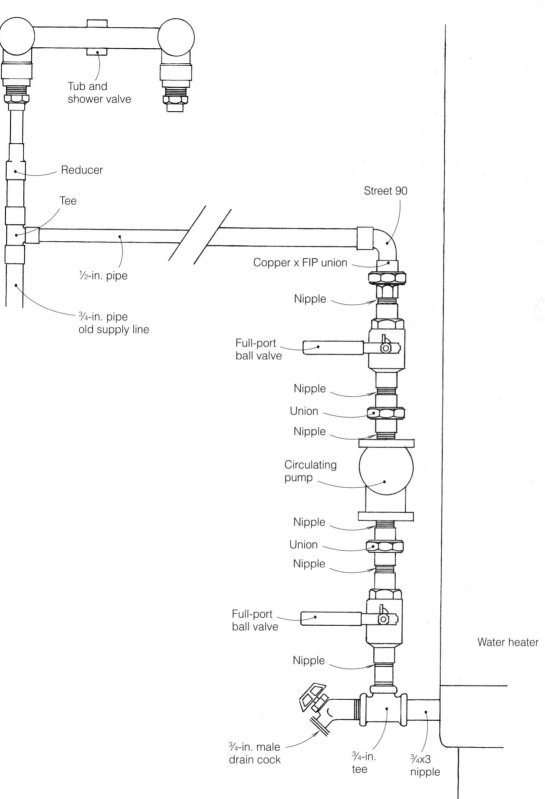

Tub and shower valve

Reducer

Tee

½-in. pipe

¾-in. pipe old supply line

Street 90

Copper x FIP union

Nipple

Full-port ball valve

Nipple

Union

Nipple

Circulating pump

Nipple

Union

Nipple

Full-port ball valve

Nipple

Water heater

¾-in. male drain cock

¾-in. tee

¾x3 nipple

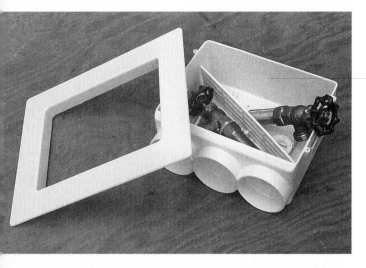

Among the options for mounting the washing-machine hot and cold supply piping are a plastic laundry box that contains the hose bibbs (top) and a single-lever laundry valve (above).

WASHING MACHINE

Hot and cold water-supply lines to the washing machine end in hose bibbs (shut-off valves), to which the washing-machine hoses attach. There are several ways to make this connection. One is to install a recessed plastic laundry box that contains both the bibbs and drain-line standpipe (if you don't use a utility sink to drain the washer into). I don't like these plastic boxes because they are cheaply made and usually end up getting cracked or broken before or during the drywalling phase. Also, there is no easy way to support the hose bibbs securely; in most cases, they end up being installed vertically, with the handles toward the back wall of the box, where they are inconvenient to operate.

Another option is to install a single-lever laundry valve. This valve, which resembles an old-style electrical knife switch, is small and convenient to operate but is impractical to repair when its seals fail. Unlike standard laundry valves (8-in. spread), this valve calls for not much space between water lines, so it's not a good choice if the pipes are installed inside a wall, since you have to open the wall to move the piping if the valve starts to leak.

A third option, which lends a touch of style, is an in-the-wall laundry valve with decorative trim. These valves, which are made by several manufacturers, are usually based on a tub/shower valve with separate, additional, copper sweat to hose adapters. Because the valve has decorative trim, it is roughed in at a comfortable operating height above the top height of the washer's control panel in clear view. Since the moving-parts portion was originally a valve designed as a tub and shower valve that would receive daily use, it is therefore much less prone to leakage. Because the valve is convenient and easy to use, you are more likely to turn off the water to the washer when it's not in use. If you shut off the water supply to your washer and then activate a cycle, just for a few seconds, using warm water (which draws both hot and cold water from the supply) and then turn the machine off again, the rubber supply hoses are relieved of pressure until the next wash and last much longer. You also spare yourself the risk of a flood from ruptured pressurized hoses.

Water-supply piping to the washing machine can terminate in hose bibbs (flanking the standpipe) mounted on a board.

HOSE BIBBS

STANDARD HOSE BIBB FOR WARM CLIMATES

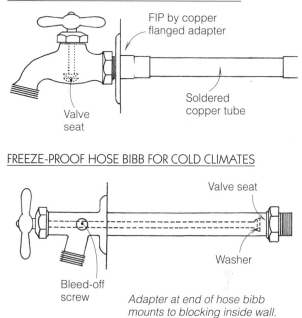

FIP by copper flanged adapter

Soldered copper tube

Valve seat

FREEZE-PROOF HOSE BIBB FOR COLD CLIMATES

Valve seat

Washer

Bleed-off screw

Adapter at end of hose bibb mounts to blocking inside wall.

A fourth, and more Spartan, option is simply hanging the bibbs on a board. Two ½-in. copper by FIP drop-eared 90° elbows inside the wall are screwed to 2x6 or bigger blocking with a rough-in height set several inches above the top height of the washer's control panel. (Most full-size washers are near 46 in. to the top of the console.) Into the 90° elbows you thread brass nipples, and onto these you thread ½-in. FIP washing-machine bibbs. At this point, to leave the smallest-diameter stub for the drywallers to fit over, just put galvanized caps on the nipples so the lines can be tested.

EXTERIOR HOSE BIBBS

The last supply piping I install is the rough-in plumbing for the hose bibbs outside the house. How many hose bibbs you install depends upon code minimums and the owner's wishes, but you can install as many as you want.

The method of anchoring the fitting for the hose bibb depends on the climate. In mild climates, you can bore a hole through the siding and screw a sill

flange (FIP by copper flanged adapter coupling) to blocking inside the wall, then solder the copper tube into the back of this fitting (½-in. dia. pipe is adequate, but ¾-in. dia. pipe is better). I usually solder a short piece of tube into the coupling first and then screw it in place in the hole, so I won't set the inside of the siding on fire when I solder on the pipe. I then trim down the pre-soldered tube with the "knuckle buster" tubing cutter (see p. 65) to a workable length once inside. Into the FIP threaded face of the coupling on the outside of the building, I thread a high-quality, long-shank hose bibb.

In cold climates, you need freeze-proof hose bibbs (see the drawing above). These bibbs have a very long shank (with the washer and seat at the end), which when installed will be inside the heated structure. There is usually a bleed-off screw on the exterior end of the bibb that allows you to drain out any water in the long shaft when temperatures turn cold. To find out the length of the hose bibb that you will need for your area, ask your local authority. One form of this freeze-proof hose bibb has male iron pipe threads on the end of the shank. You still use an eared FIP by copper fitting that is anchored at the required distance inside the building, into which you thread the hose bibb.

The hose bibbs themselves are installed later, after the water test. For now you simply thread a galvanized nipple with a cap on the end into the fitting for test purposes.

TESTING THE WATER SUPPLY

Water-supply testing of a building with an off-the-ground foundation happens when the framing is completed. As with the DWV system (see pp. 128-130), the usual test is a water test. The plumber installs the entire fresh-water system with the outlets capped off and then uses a neighbor's water (via garden hose) to charge the lines with the actual water pressure of the neighborhood. Water systems may also be tested with compressed air, but this is a more difficult procedure to set up, and it is harder to locate leaks.

If your supply system is polybutylene, the testing procedure is somewhat different, as described below.

TESTING A PB SYSTEM   It is best to test a polybutylene system with water, if it is new construction and only plywood and stud wall would get wet in the event of a leak. Most pipe manufacturers recommend using water in a hydrostatic test at 200 psi. You can purchase a hand-operated hydrostatic test pump, the Wheeler-Rex #29100, for less than $200. If you are doing an upper-story addition or remodel and any possible leak would wreak havoc with furnishings, you could air test. Here, you introduce about 100 psi into the system and watch the gauge. If you have a leak, you need to locate it with the aid of a spray bottle of soapy water; look for bubbles produced by the leak. PB pipe manufacturers frown on air tests because they fear someone might use a detergent that would damage the pipe. Rex Cauldwell, an avid PB proponent and plumber in Copper Hill, Virginia, says that his supplier of PB approves his application of a 2% solution of Palmolive Green for this purpose.

SOLAR PREP

There are so many different solar heating systems that it would be impossible here to give details on them all. So I'll keep my comments general. Now, while the walls are still in stud, is the time to add a loop for solar heating. Find out what diameter piping would be required for a system best suited for your building and the best location to start and end the loop. You will need to pressure-test this loop independently of the rest of the fresh-water system for the house.

Another possible solar installation is the solar still, which uses evaporation to purify poor-quality fresh water. You must have a level spot to set it on. If you had a steeply pitched roof, but with maybe an area with good sun that isn't too visible, you could provide permanent, custom-built roof jacks to set one. What is important now is running a loop for the still before the walls and ceilings are in. The still is filled daily by a generic irrigation timed valve, and the evaporated water that is collected runs by gravity to a tank (usually in a kitchen cabinet or the attic above and near the sink). From the tank down to a separate faucet at the sink you run a small-diameter ($\frac{3}{8}$-in. coil) line. The incoming water on each cycle flushes out the residue of poor water into another gravity drain, which you bring to a rain gutter or all the way down to the ground. The valved line is the only line that must get pressure tested, but it is a good idea to test all three lines under pressure before they are encased in the walls.

HOT-WATER HEATING SYSTEMS

In some parts of the country, it is common practice to heat the home with a hot-water boiler and radiator system and generate the domestic fresh hot water using either a separate boiler or a heat exchanger in conjunction with the boiler. The two systems are sealed and never mix. If this type of system is in your future, I recommend that you get factory-supplied schematics for water supply locations now, so you won't have to chop out and move supplies to the proper location for your particular unit later.

Hot-water heat is a good system, because the warmed air is not robbed of moisture. But it is an expensive system; it uses a sophisticated boiler and lots of ¾-in. copper circulating lines. You might be able to install these circulating lines yourself to save on the total costs instead of having a hydronic engineer install them. Once installed, they have to be tested, and you will need to do that before sealing any of them up in the walls, floors or ceilings. On a slab job it would be tempting to put these lines also under the mud, but I recommend that you do not because the lime in the slab will destroy the copper and problems will be costly to fix.

Another type of hot-water heat is the radiant slab. When this technology was first being tried about 30 years ago, copper piping was laid right into the slab. The slab ate the pipe and the systems had to be abandoned. Today, CPVC plastic piping is used with standard concrete, copper is used with limeless concrete, and the systems are quite workable, albeit expensive. Don't attempt to design or install a radiant system yourself; consult with someone who has experience designing, installing and maintaining these types of systems.

WATER SUPPLY WITH A SLAB FOUNDATION

On a slab foundation, after the DWV piping has been installed (see pp. 131-136), you can turn your attention to the water-supply lines; these too must be installed before the slab is poured. You have three possible courses of action. The first, which I recommend, is to run a properly sized building service line (usually ¾ in. or 1 in. in diameter) of locally acceptable materials above or alongside the slab; from there, the supply and distribution piping can be strung in the walls and other convenient areas just as for off-the-ground construction. The second option is to lay a main supply under the slab that is brought up out of the slab in a central portion of the house, with hot and cold branch lines strung in the stud wall as for off-the-ground construction. The third choice is to lay most of the supply and distribution system under the slab, letting it surface in the walls, with short runs to the fixtures. Burying most of the pipe under the slab appeals to builders because they save the time and labor costs of boring a lot of holes and stringing pipe above slab. Only this third option need be discussed here.

SUPPLY PIPES UNDER A SLAB

With a slab foundation, the plumber needs to get in and out so the job can progress; all the work is held up until the pipes are staked in position so the slab can be poured. So once the under-slab part of the DWV piping is done, you get the under-slab portion of the water supply and distribution piping installed, tested and signed off on, then return when the house is framed to finish the rest.

All codes want as few joints as possible in water pipes buried under a slab, and the requirements governing these joints are especially stringent. With rolled copper, most codes will accept only wrought-copper water fittings (no wrought brass) and require any joint to be brazed with a silver-based or other high-strength alloy filler metal. In most cases the pipe will need to be sheathed in a plastic sleeve to protect it from the lime in the concrete, which leaches out with moisture and travels in the soil directly below and adjacent to the slab. Minerals will destroy almost any metallic water-piping material, and they are especially corrosive to copper. Having had to jackhammer slab floors searching for leaking buried water piping, I strongly recommend that you consider all other pathways before committing any fresh water to under-slab burial.

For installing any piping under the future slab, use the double string lines denoting plumbing-wall locations that you set up to lay out the DWV pipes (see the drawing on p. 133). Remember that in warm climates you'll need a through-form sleeve in the foundation wall to provide an avenue for the copper or other pipe to pass through from the outside if it isn't a safe distance under the perimeter footing. One such product is Sure-Sleeves, made by Specialty Products; you could also use a length of large-diameter Schedule 40 PVC pipe.

The farther below the slab you can lay your copper water lines, the better, because if the sheathing deteriorates, the lime that leaches out of the slab will eat away the copper. Most codes do not allow water pipes to be run in the same trench as building sewer or drainage piping constructed of clay or materials that are not approved for use within a building, such as terra-cotta. So with ABS, PVC or no-hub iron under the slab (and the inspector's approval), you might be able to lay the water pipe in the same trench as the sewer or drainage pipe, if the trench happens to be going in the same direction that you want to send the water supply in.

The deeper you go below the slab with your water, the more heat your hot-water lines can lose to the soil, so you might consider thermal insulation (now is the only convenient time to install insulation). Ask your local building inspector and plumbing supplier what form of insulation best resists rotting in your soil. In most cases, well protected (sheathed and taped) copper water piping can be brought up close to the gravel bed under the slab, and radiant heat from the building will protect the supply from freezing.

If I were dealing with a below-slab water line with branches to several different locations, involving the installation of tees or other joints that would be buried under the slab, I would run entire additional lines of sanctioned material instead from the water source to the fixture locations, even though that would use a lot more piping. If the pipes were rolled Type K copper, the cost of the material would be very high, but believe me, it would be worth it. If the pipes could be PB, your wallet would breathe a whole lot easier. If the day ever comes that you also have to jackhammer up someone's living room or kitchen floor to get at a bad joint, you will comprehend this apparently extravagant precaution of using entire jointless lines.

Also, as with the DWV stub-ups, when you bring the pipes up from the slab make sure that they are plumb, because framing crews are not known for patience. If the stub-ups are straight, the framers can drop their bored plates down over them with ease. If the stub-ups are considerably out of plumb or bent over, the framers might chop them down some to get their plates in place. More realistically, they might cut their plates and use individual pieces of material between pipe stub-ups, and the inspector might make you install reinforcing metal FHA straps between the severed plates. If the stubs have been cut or tampered with, joining to them later can be a real headache.

Supply-pipe stub-ups should have about 6 in. to 8 in. of material above finish-floor height. It's a good idea to mark the piping hot and cold and from where the piping was brought and where it's going, using either a permanent marker or a wired-on tag. For both the copper and plastic piping you might want to wire on tags that warn of impact damage and ask people to keep building materials from direct contact with the piping.

MANIFOLDS

If your project has several bathrooms, a kitchen and a laundry, and you decide to run individual, branchless pipes to each, then you might have four or five individual hot-water lines leaving the area of the future water heater. After the slab is poured, these lines can be manifolded (joined into one short, bigger pipe with tee branches). For now, you can just hook up the lines (using 90° elbows) leaving the water heater's location. As shown in the top drawing on the facing page, each line heads off to its respective room, where it is stubbed up and capped. (For two-story construction, the lines are stubbed up out of the slab at locations affording convenient continuation to the upper story.)

Cold-water supply lines can be treated similarly. You can run individual lines from the supply at one edge of the house (manifolded as at the water heater)

MANIFOLDS: PRELIMINARY HOOKUP ON SLAB

OVERHEAD VIEW

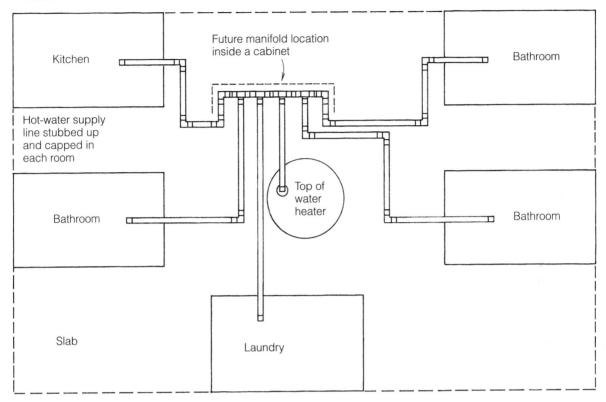

Kitchen

Future manifold location inside a cabinet

Bathroom

Hot-water supply line stubbed up and capped in each room

Bathroom

Top of water heater

Bathroom

Slab

Laundry

MANIFOLD FOR HOT-WATER LINES

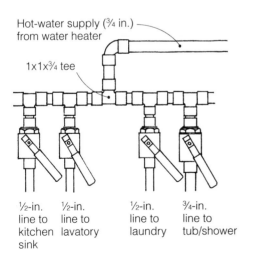

Hot-water supply (¾ in.) from water heater

1x1x¾ tee

½-in. line to kitchen sink

½-in. line to lavatory

½-in. line to laundry

¾-in. line to tub/shower

All valves are full-port ball valves for balancing pressure drops; supply-line sizes are typical.

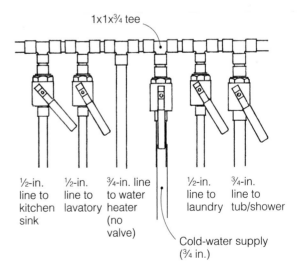

1x1x¾ tee

½-in. line to kitchen sink

½-in. line to lavatory

¾-in. line to water heater (no valve)

½-in. line to laundry

¾-in. line to tub/shower

Cold-water supply (¾ in.)

to their respective rooms. Or you can bring in one main cold supply to whatever room you want (possibly the utility room) and then pull additional lines from the remaining wet rooms over to the main. Using tees, you can temporarily join each individual line to the main for the first-floor slab test and permanently manifold the lines later. It is best to do this manifolding in an accessible location.

On each individual water line leaving the manifold you'd be wise to install a full-port ball valve, as shown in the drawing on p. 161. This is so you can throttle down (adjust) the flow to balance out pressure drops that result when other fixtures in the house are used. Generally you want a lot of water flowing to the washing machine and the shower, less to sinks and lavatories. That way a person taking a shower won't get scalded or frozen when someone in the kitchen turns on the tap.

TESTING THE BELOW-SLAB WATER SUPPLY

As with the DWV system, the below-slab supply system has to be capped off, water tested or air tested and approved by the inspector before the slab can be poured. PB pipes are capped off with a barbed by MIP adapter to which a cap can be threaded. If you are using rolled copper, don't just crimp the edge closed and solder it; use a soldered cap. Crimping can push the top inch or so of the pipe out of round, and if the stub is already on the short side, you might have trouble later getting a good soldered joint at a tee or coupling attached at this point.

For a water test, attach a male iron pipe adapter (MIP) to the outside of the hose-bibb supply line. This male adapter is made in each piping material. To this adapter you can thread a FIP to female hose

TESTING SUPPLY PIPES ON A SLAB FOUNDATION

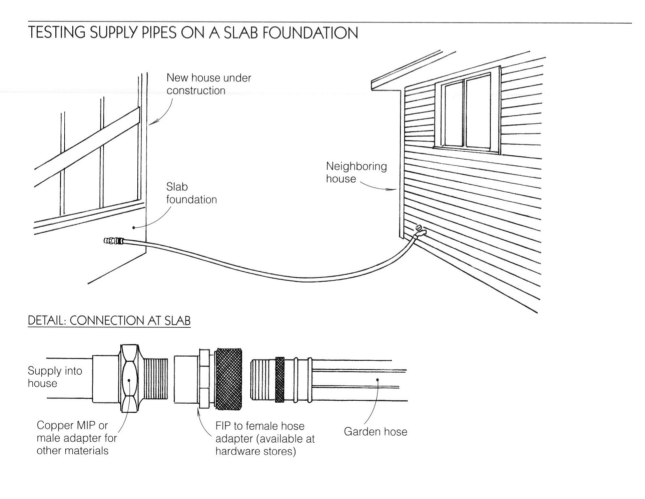

New house under construction

Slab foundation

Neighboring house

DETAIL: CONNECTION AT SLAB

Supply into house

Copper MIP or male adapter for other materials

FIP to female hose adapter (available at hardware stores)

Garden hose

adapter and then attach a garden hose. The garden hose can be hooked up to an already existing meter for this job site, or you can borrow the neighbor's.

If there is no convenient water source on the job site yet, you can test the water lines with air. Most local authorities call for a 50-psi air-pressure test on the line. The appropriate gauge for water-supply testing may be purchased at your local plumbing supplier. Make sure to test the air fitting and male threads of the gauge when it's under pressure. Where the gauge and the air fitting thread into the body is often leaky, even on brand-new gauges. If the gauge leaks, remove the part, wrap it in Teflon tape, coat the threaded female opening with pipe dope and reassemble.

Once you have checked the supply pipes for leaks, they must be wrapped with an acceptable isolating material wherever they will be touching or passing through concrete. Multiple layers of PVC gas-wrap tape, plumber's foam wrap and extruded closed-cell foam insulation are all possible choices. Check with your local authority as to which is acceptable for the piping you are using.

If the pipe runs in the slab more than 5 ft. or 6 ft., it might be necessary to anchor the pipe in position to prevent it from floating up when the slab is poured. Resist using form wire for this job because it can cut the insulation when the mud is trying to float it. Use the wider galvanized or plastic plumber's tape (see pp. 35-36) wired to driven metal stakes. If you have black gas-wrap tape or traditional duct tape (silver grey) wrapped for several layers starting 5 in. or 6 in. back from each side of the insulation joints or tee-branch intersections, it will seal out the moisture long enough for the concrete to set.

With the piping so wrapped, you can call for an inspection on the below-slab rough plumbing. Once the inspector signs off, the plumber gets a vacation from the job until the framers have the stud walls and maybe even trusses or rafters in place. Then the lines are uncapped and the supply pipes above the slab can be installed and tested, just as with an off-the-ground construction.

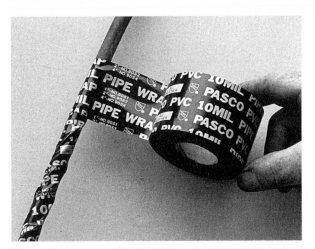

Wrapping copper tubing in PVC gas-wrap tape is a code-approved way to protect it from the corrosive effects of burial in a concrete slab. The tape also allows room for the pipe to expand and contract.

CHAPTER 7

FUEL-GAS DISTRIBUTION PIPING

This chapter deals with pipes that carry fuel gas, which can be either natural gas or liquefied petroleum (LP) gas, commonly referred to as propane. In some communities you may not have a choice of gases; in other areas you will find both fuel options. Gas distribution piping conveys the fuel gas from a gas main or storage tank into the house to any appliances that require it — most commonly the kitchen range, the water heater, the clothes dryer and/or the furnace. Gas vent piping, which carries the exhaust to the outside, is discussed in Chapter 8.

Gas distribution piping is a pressurized system; the pressure is supplied either by the natural-gas utility as it pumps the gas to your neighborhood in the street mains or by a pressurized propane tank. The gas utility is responsible for the installation of the building's service pipe from the source of supply to the structure. In my area, only natural gas is used, and for that I install threaded-steel piping. However, the materials and methods of installing a threaded-steel distribution piping system are the same for either gas. (Additional piping materials are sanctioned in other areas; these are discussed on p. 165.)

LP gas is a by-product of making gasoline. It is stored in a tank that is hauled to your house on a truck and hooked up to your fuel distribution line. When the tank is empty, the truck comes back to refill it or bring you another. Because LP gas is stored in a tank until you use it, it is very useful to people in rural areas and areas not served by a natural-gas pipeline. Natural gas, on the other hand, travels from the well (source) to you in a pipeline, and you need only turn a valve to let it in the house.

Gas is a dangerous substance, and local codes may not allow plumbers to install fuel-gas lines. If you cannot legally install these systems, secure the services of a specialist who can (check the Yellow Pages under "Gas — Liquefied Petroleum").

LP gas is much more dangerous than natural gas. In its storage tank, LP gas is a liquid because it has been compressed. When pressure is released, as it passes through the piping, it turns back into a gas. However, the gas is heavier than air. When it leaks into the atmosphere (or into a room), it seeks the lowest point and collects. There is no escape for this collected gas until it is finally absorbed by the atmosphere, which can take several hours. Collected gas is extremely dangerous. If the level builds up to the height of a pilot light on a water heater, stove or oven, the gas will ignite and explode. You should think about this when you are working with LP gas piping. Collected gas can be dispersed with a ventilating fan, but make sure that the fan you use meets stringent anti-sparking standards (look for one with a classification of OSHA/ANSI Group D). In my area, anti-sparking fans rent for about $150 per day.

PIPE MATERIALS, FITTINGS AND VALVES

Fuel-gas piping has its own set of requirements for materials, fittings and valves. Some of these are stipulated by major or local codes; others are a matter of individual preference.

MATERIALS

Many materials are used for gas piping, and you should check your local code to see what is allowed in your area. The only materials I am allowed to use for fuel-gas distribution where I live are threaded steel (black, galvanized and protectively coated) and yellow brass. But because yellow brass is prohibitively expensive, no plumber I know uses it. In some areas Schedule 80 PVC can be used for fuel gas; again, check with your local authority.

In some parts of the country, untreated (clean) Type K rolled copper tubing is allowed for LP gas. Untreated copper is not allowed in my area for natural gas because it reacts with the sulfur in natural gas and forms a black feathery soot, which eventually clogs the pipe completely. In some other areas governed by my major code, natural gas can be conveyed in tinned copper.

In locales where it is permitted, you might find entire gas distribution piping systems in copper with flare fittings. But I feel safer seeing a home's distribution piping done with traditional threaded-steel piping because it is more capable of withstanding the abuse some home owners subject exposed portions of their systems to.

I use black threaded-steel pipe inside the building, more for identification purposes than for cost savings. It's actually the easiest material to cut and thread. If I need to run gas on the outside of a building, I use galvanized-steel pipe and run it no closer than 6 in. to grade. For burying in the ground, I am required by code to use the protectively coated black or galvanized steel.

FITTINGS

For threaded-steel systems, I use galvanized fittings only. (There is no such thing as protectively coated fittings, so I use galvanized here too and then hand wrap the exposed section of pipe, including the galvanized fitting, prior to backfilling.) I have learned from experience to stay away from black fittings: Too many of them have little pinholes that will leak. If you have a bad fitting, your piping will never hold enough air-test pressure (10 psi to 30 psi) to pass inspection, and could possibly even leak gas at the lower gas operating pressure, which is in ounces. Galvanized fittings (shown in the photo below) start out just like black ones. But after being cast, they are hot dipped in molten zinc, which seals the holes. Try to inspect every fitting before using it. Look closely for hairline cracks and small holes. Also look for deformities in the threads.

Black steel pipe with galvanized fittings is a good choice for gas supply piping inside a building.

Merchant's couplings (left), which come on the end of the pipe, should be replaced with forged couplings (right), which have a thicker wall and are less subject to stretching.

When you need to couple one piece of threaded-steel pipe or nipple to another, do not use the coupling supplied that comes on the end of the pipe when you buy it. These couplings, called merchant's couplings, are thin-walled and have a tendency to stretch when the male is tightly installed, causing a leak. It is much safer to use forged couplings, which have a thicker wall with a raised welt around each end that further protects against stretching.

Instead of a coupling, I occasionally use a tee, with branch down and plugged, when my horse sense tells me to do it. Since the lines I install are always oversized (see pp. 169-171), these tees can be used in future modifications, or the plugs can be replaced with drip-pipe nipples and caps (see p. 168).

When installing a copper piping system, you should check with your local authority for specifics. Flare fittings are used with copper to transmit fuel gas (see pp. 72-73). When installing plastic systems, check with your local authority first for any special conditions, especially involving any forbidden fittings or specified fittings for use with fuel gas.

VALVES AND REGULATORS

The LP gas in your storage tank may be under a pressure as high as 150 psi to 175 psi. (In contrast, natural gas in the street main might be at 20 psi to 25 psi.) With LP gas, there are various regulators to reduce this pressure before it reaches your appliances. Usually a regulator at the tank brings down the sending pressure to around 10 psi. Once the line gets to your structure, another regulator reduces the pressure even further, down to about 11 in. of mercury or a mere 2 oz. to 3 oz. per sq. in. for the piping system within the house.

Shut-off valves used for fuel gas (see the top left photo on the facing page) may have packings that are different materials than those used just for water. Use only shut-off valves listed for the particular medium you are working with. Many manufacturers produce high-quality ball valves that have the letters WOG cast on the valve body. WOG stands for "water, oil and gas," and this valve may be used with piping that conveys any of the three. WOG valves usually have a pressure rating of 125 psi to 150 psi.

Gas only, in-line ball valves also are constructed to withstand fairly high pressures, even though they are subjected only to ounces of working pressure and maybe as much as 30 psi on an air test. Their handles are quite small compared to a WOG valve, so they turn very freely. This is a nice feature when you wish to interrupt service. Most of these valves have female iron pipe threads. In-line gas ball valves with standard female iron-pipe threads can be used with copper by using MIP by flare adapters threaded into the valve on each port.

I would advise you not to use a tapered-plug valve (see the top right photo on the facing page). I have broken the handles off countless tapered-plug valves when trying to turn them off because they are difficult to operate and the handle is too thin at the base. Also, many in-line tapered-plug gas valves will not withstand much more than 10 psi. Your system will probably fail an air-pressure test at 20 psi and higher, and some inspectors may ask for 30 psi. Ball valves designed for fuel gas can easily withstand these higher pressures. Ball valves are so superior to plug valves that it pays to spend the extra money for them and leave your problems at the supplier.

Shut-off valves suitable for use on fuel-gas lines, shown in the photo at left, include the gas only, in-line ball valve (left) and the WOG ball valve (right). The tapered-plug valve, shown in the photo at right, should be avoided.

HANGING FUEL-GAS PIPE

The two-hole, galvanized-steel strap hanger is still widely used for securing fuel-gas pipes to wooden structural members. If you have a two-story house with a basement or a crawl space, you can hang the piping from the bottom of the first floor's floor joists, just as you did the water-supply pipes (see p. 139). When hanging threaded-steel fuel-gas lines below structural members, the task is usually best accomplished by bolting a single strand of galvanized plumber's tape to the pipe and then securing the tape to perpendicular or parallel floor-joist bottoms or to a 2x4 or 2x6 drop support when swing fitting or crossing joists diagonally. When done this way, the installation is strong, although it is a bit unsightly.

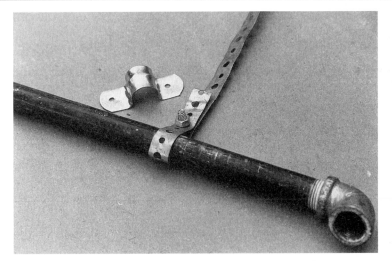

Threaded-steel fuel-gas lines can be supported with a two-hole, galvanized-steel strap hanger (left) or galvanized plumber's tape (right).

Another option is to use uncoated wire hangers (see p. 38). For supporting vertical threaded-steel gas lines in pipe chases, you can bolt lengths of galvanized plumber's tape below convenient couplings, tees or 90° elbows and then secure the tape to wood structural members overhead.

If you have the choice of plated-steel or solid-brass valves, buy the brass ones because they seal much more efficiently to the pipe. Shut-off valves to be used with flexible appliance connectors at the fixture can be purchased with male flared discharge ends. The flared ends save you from having to use an additional nipple and separate flare adapter.

If your fireplace will have a log lighter under the fireplace grate, you'll have to install a log-lighter valve for turning the gas on and off (for this you use a square-drive-log-lighter key). The valve may be installed in the floor or wall, and the valve's trim is usually installed flush with the floor or the outside surface of the masonry firebox, respectively. My major code says that the log-lighter valve may be no more than 4 ft. away from the fireplace opening.

On the other side of the valve you continue on through to the inside of the firebox. This second pipe run from the discharge side of the log-lighter valve may be very short, but you'll have to pressure test it. If you leave the valve out and run the line to the firebox, you will have to come back and splice the valve in after the pressure test with a right/left coupling (see pp. 173-174). Because the log lighter often isn't installed until after the rest of the fuel-gas system is tested, it will need its own separate test. If your local authority will allow you a separate 10-psi test, then you can install the log-lighter valve now, when you are doing all the gas lines, for it will pass. This will save you a lot of time.

DRIP PIPES

Drip pipes are a trap for debris in the gas pipe — rust particles, dust or water — and my major code calls for their use when water vapor is present in the fuel gas. Drip pipes also protect sensitive gas-regulating devices from any rust and scale that gets into the system. How long should they be? It often depends on local superstitions. If every plumber since your great grandfather has been adding 6-in. or 8-in. drip pipes in your locale and every inspector since then has gotten used to seeing them that length, that's how long they should be.

Whether the gas supply is vertical or horizontal, drip pipes are installed on the bottom inlet of a tee (see the drawing below). The incoming gas passes through the tee and out to the gas-control valve of the appliance. Because of its weight, any debris in the line will fall straight down to the capped bottom nipple. In some locales, water-heater manufacturers may void the warranty on their units unless drip pipes are installed.

DRIP PIPES

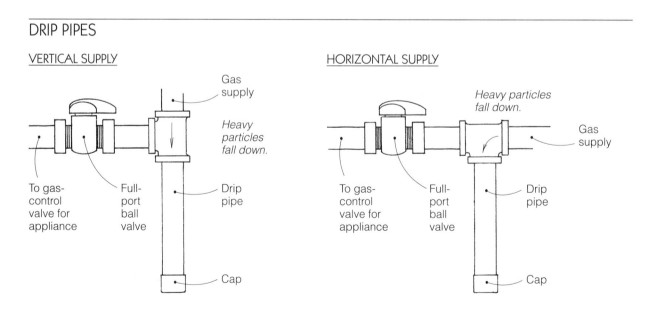

As gas passes through the pipes, any debris in the line falls straight down the drip tube, where it won't clog valves.

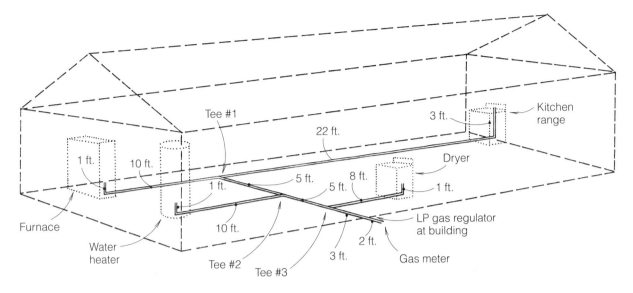

SIZING PIPE FOR GAS

Sizing the fuel-gas system is somewhat easier than sizing the water-distribution system; the approach is similar, but there are not many gas-fired appliances to deal with and the charts are easier to read. You first sketch the proposed system, noting the distances between the fuel-gas source and each appliance. You also need to ask your gas supplier the pressure at which the gas will be operating inside your piping system. Then you note the gas demand of each appliance. My major code has a table that lists each appliance by its demand for gas at various pressures. Other tables specify how much fuel gas is conveyed through various lengths of pipe at different diameters.

The drawing above shows a hypothetical natural-gas piping system with an operating pressure of 2 psi. Let's work through the steps involved in sizing this system. My code instructs me to start with the appliance farthest from the source (the meter or the regulator at your house) and measure back to the source, then find the pipe size for this distance at this operating pressure. As shown in the drawing, it's 25 ft. from the kitchen range to tee #1 and 15 ft. from tee #1 to the gas meter — a 40-ft. run in all.

TEE FITTINGS AT BRANCHES

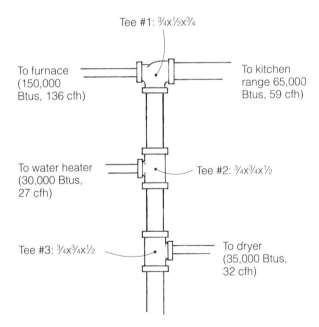

MINIMUM DEMAND OF TYPICAL DOMESTIC GAS APPLIANCES

	DEMAND (BTU/HR)	DEMAND (WATTS)
RANGE	65,000	19,045
RECESSED TOP BURNER RANGE SECTION	40,000	11,720
RECESSED OVEN RANGE SECTION	25,000	7,325
WATER HEATER (UP TO 30-GAL. TANK)	30,000	8,790
WATER HEATER (40- TO 50-GAL. TANK)	50,000	14,650
DRYER	35,000	10,255
FIREPLACE LOG LIGHTER	25,000	7,325
BARBECUE	50,000	14,650
REFRIGERATOR	3,000	879
FURNACE	150,000	43,950

This chart is adapted from UPC "Fuel Gas Piping."

NATURAL GAS: MAXIMUM CAPACITY OF PIPE (CUBIC FEET PER HOUR)

PIPE SIZE (IN.)	LENGTH (FT.)											
	50	100	150	200	250	300	350	400	450	500	550	600
½	466	320	257	220	195	177	163	151	142	134	127	121
¾	974	669	537	460	408	369	340	316	297	280	266	254
1	1,834	1,261	1,012	866	768	696	640	595	559	528	501	478
1¼	3,766	2,588	2,078	1,779	1,577	1,429	1,314	1,223	1,147	1,084	1,029	982

This chart is adapted from UPC "Fuel Gas Piping." The chart assumes a gas pressure of 2 psi with a drop to 1.5 psi.

LP GAS: MAXIMUM CAPACITY OF PIPE (THOUSANDS OF BTU PER HOUR)

PIPE SIZE (IN.)	LENGTH (FT.)											
	10	20	30	40	50	60	70	80	90	100	125	150	200
½	275	189	152	129	114	103	96	89	83	78	69	63	55
¾	567	393	315	267	237	217	196	185	173	162	146	132	112
1	1,071	732	590	504	448	409	378	346	322	307	275	252	213
1¼	2,205	1,496	1,212	1,039	913	834	771	724	677	630	567	511	440

This chart is adapted from UPC "Fuel Gas Piping" and based on a pressure drop of 0.5 in. water column; low-pressure 11-in. water column.

In the chart at the top of this page, we see that the kitchen range has a Btu/hour demand of 65,000. Since the sizing tables for natural gas are listed in cubic feet per hour (cfh), you have to convert the Btus to cubic feet. To do this, divide the Btu figure by 1,100. The kitchen range, therefore, has a demand of 59 cfh. Now look at the middle chart, which gives pipe diameters for various lengths of run and pipe capacities. Since the first column (50 ft.) is greater than our pipe run of 40 ft., we use this column. Reading down, we find that ½-in. dia. pipe will supply 466 cfh, more than enough for the kitchen range, so the diameter of that pipe should be ½ in.

In the example (see the drawing on p. 169), three other appliances share the pipe run from the meter: a furnace, a water heater and a dryer. We go through the process for each appliance, beginning with the appliance next farthest from the gas meter, which in this case is the furnace. The furnace is 26 ft. from the gas meter and the appliance demand is 136 cfh (150,000 Btus ÷ 1,100). At tee #1 the combined demand of the kitchen range and the furnace is 195 cfh, well within the capacity of ½-in. pipe. According to cfh demand, we know we can use ½-in. pipe from furnace to kitchen range and at least down to tee #2.

The next appliance, the water heater, is 21 ft. from the gas meter. It has a demand of 27 cfh (30,000 Btus ÷ 1,100), yielding a combined appliance demand thus far of 222 cfh, still well within the capacity of ½-in. pipe. Therefore, according to cfh demand, we can use ½-in. pipe at least down to tee #3.

The last appliance, the dryer, is 14 ft. from the gas meter. It has a demand of 32 cfh (35,000 Btus ÷ 1,100), yielding a total cfh demand for the house of 254 cfh, still way under the limit for ½-in. pipe at 5 ft. from the source.

The total footage distance for the system (codes refer to this distance as developed length) is 71 ft. For this length of pipe, the middle chart on the facing page tells us we could create the entire system in ½-in. pipe. However, cfh is not the only factor to consider. Many local codes stipulate that all freestanding ranges must have ¾-in. dia. supply piping, regardless of their cfh. This means that from the range back to tee #1 and from tee #1 back to the meter, the piping must be ¾ in. But, if the range were any other appliance, you would initially size the system just as we did here. However, I would also recommend that you increase the diameter of each pipe required in your system by one pipe size to accommodate possible additions to the system in the future.

In this exercise, as per code minimum, the final tee configurations would be as follows: tee #1 would be ¾x½x¾, tee #2 would be ¾x¾x½ and tee #3 would also be ¾x¾x½, as shown in the detail drawing on p. 169. The pipe runs from the furnace to tee #1, from the water heater to tee #2 and from the dryer to tee #3 could still be done in ½-in. pipe.

If you are sizing a distribution system for LP gas, you can work through the calculations again, this time referring to the bottom chart on the facing page instead of the middle chart.

LAYOUT CONSIDERATIONS

Major codes do not permit the burial of fuel-gas lines under slab, so both off-the-ground and slab houses have to be piped for fuel gas through wood members. Your code may also have regulations about the placement of the gas meter, but the gas utility or supplier usually has the final say. For LP gas regulators, the local authority might take a greater interest.

Codes vary greatly, but it is a good idea to keep gas piping several feet away from any heat duct, ventilation vent, appliance vent, chimney or electrical panel. Use as few couplings as possible. Threading pipe takes a lot of time, and the more joints, the greater the risk of leaks. If you have a single-story or split-level structure with a low attic and standard drywall ceilings, it is easiest to bring the main gas line into the house vertically in an exterior wall and then branch off with runs through the attic, dropping down again in the walls backing or hosting the appliances. If you are lucky, you can do all of the piping this way.

Sometimes, as with fresh-water supply piping, you have to send the piping through long horizontal distances of stud wall. If you do, you might be able to avoid the need for tedious couplings by boring a hole through an exterior wall and sliding in a long pipe (see p. 140). If that isn't possible, you might ask if your inspector has any objections to notching a run of studs or ceiling joists. Notching is a structural issue in some instances, and you may need to discuss the location and depth of the notches with the designer or architect.

JOINING THREADED-STEEL FUEL LINES

Generally speaking, threaded-steel fuel-gas lines are joined similarly to threaded-steel fresh-water distribution lines (see pp. 84-86), with the exception of three types of joints: right/left couplings and nipples, ground joint unions, and increasing and reducing bushings (see the photos below).

However, depending upon your locale, there may be some restrictions on a few particular threaded fittings used with threaded-steel fuel-gas lines. In my major code, inside any concealed space, only right/left couplings and nipples are accepted as unions. Ground joint unions may be used only at exposed fixture, appliance or equipment connections and in exposed exterior locations immediately on the discharge side of a building shut-off valve. Increasing and reducing bushings are not allowed in any concealed spaces; bell reducers have to be used instead. For branch lines that intersect with a larger supply line, tees may be purchased with reduced inlets and branch inlets for almost any pipe size, which can save the cost of adding nipples and bell reducers.

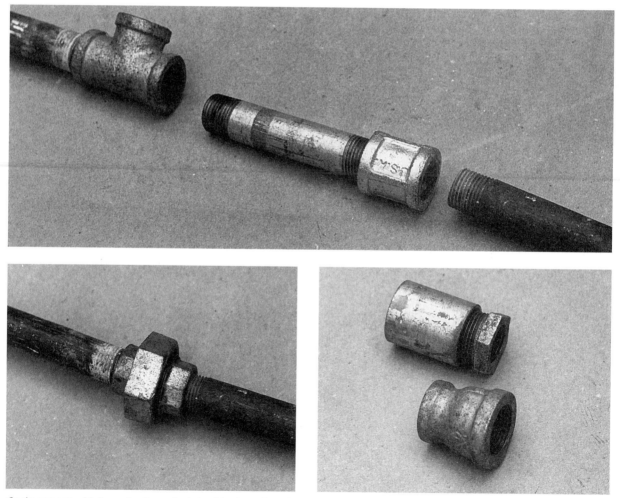

Code-accepted joinery for threaded-steel fuel-gas lines in concealed spaces is a right/left coupling and nipple (top). Ground joint unions (above left) and reducing bushings (at top in the photo above right) are not allowed in concealed spaces; bell reducers (at bottom in the photo above right) must be used instead.

REPAIRING WITH RIGHT/LEFT COUPLINGS AND NIPPLES

Most of the time leaks in threaded-steel fuel-gas lines occur at a joint, but sometimes there is a hole in a fitting or a crack in the pipe. These leaks are difficult to fix because the pipe was joined by starting at one end and continuing in one direction, so there is no way to undo the middle sections without unthreading everything that comes before it. A quicker, but more painful approach is to cut out a section of pipe run with a reciprocating saw and rethread the stub of the old pipe in place to receive a right/left coupling and nipple. However, this can be agonizing work, especially if the pipe diameter is larger than 1 in. It helps if you have a big (24-in.) pipe wrench to hold one side of the joint steady while you thread new threads with the other arm using the die head on the pipe threader (see pp. 82-83). You could also use a smaller (18-in.) wrench, and a "cheater" (a 3-ft. section of 1¼-in. pipe slid down over the wrench handle) for additional leverage.

Right/left nipples are available in two lengths: 4 in. and 6 in. Take whichever your supplier has, and be grateful if you find either. Get your right/left coupling and nipple before making any cuts. It's useful to assemble them just with your fingers, and then lay the assembly on the existing run and make your marks for cuts. The right/left nipple and coupling have knurled marks around the ends that are left-hand threads. You will need to take into acccount the number of eventual threads that will be in mesh in each side of each fitting. This distance varies with the diameter of the pipe. I have found that factory-threaded ½-in. pipe torqued into ½-in. fittings penetrates about ½ in.; ¾-in. pipe penetrates almost ¾ in.; 1-in. pipe penetrates about ¾ in.; and 1¼ in., 1½ in. and 2-in. pipe penetrate nearly 1 in.

You also have to take into account the width or depth of the ratchet and die. The longer the handle on the ratchet, the more leverage and the easier it will be to cut new threads. My Ridgid 12R is the larger of

REPAIRING A GAS LINE WITH A RIGHT/LEFT COUPLING AND NIPPLE

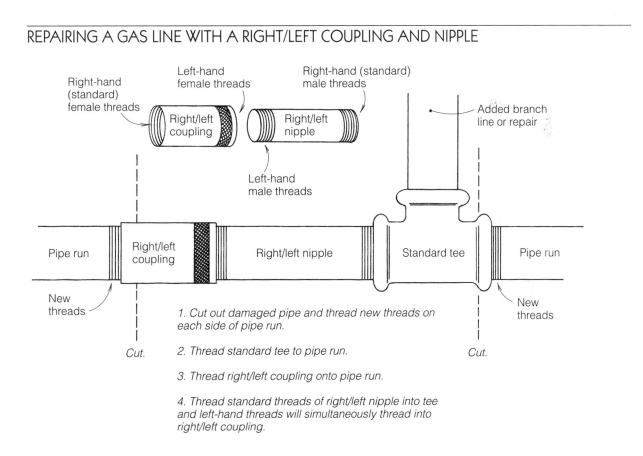

Right-hand (standard) female threads

Left-hand female threads

Right-hand (standard) male threads

Right/left coupling

Right/left nipple

Left-hand male threads

Added branch line or repair

Pipe run

Right/left coupling

Right/left nipple

Standard tee

Pipe run

New threads

New threads

Cut.

Cut.

1. Cut out damaged pipe and thread new threads on each side of pipe run.

2. Thread standard tee to pipe run.

3. Thread right/left coupling onto pipe run.

4. Thread standard threads of right/left nipple into tee and left-hand threads will simultaneously thread into right/left coupling.

my two sets and has the longer handle, so I use this tool whenever space permits. The 12R, with die in head, is almost 4¼ in. wide. The total movement of the ratchet handle from one side of the a wall opening to the other might give me only an eighth of a turn on the pipe, and I have to cut 10 to 12 threads for each end of pipe. That's about 100 hard, straining arm movements for most of these repairs.

With the bad piece of pipe cut out and the new threads cut, it's time to install the new coupling and nipple (you can also install a branch fitting at this time). Installing right/left couplings and nipples requires patience. You need to be able to push one of the two sections being joined far enough apart to get the standard right/left nipple and coupling in between the newly cut threads on the pipe run and the female threads of the standard fitting on the other side. But you still want the compression force of the pipe run to push on both ends of the right/left nipple combination. As shown in the drawing on p. 173, begin by threading the standard threads of the right/left nipple tightly into one side of the replaced fitting to the full depth. Then turn the right/left couplings's standard threads onto the other newly cut standard threads of the pipe run as you would asemble any pipe and fitting. The other side of the right/left coupling, which has the left-hand threads, will automatically draw itself onto the left-hand threads of the right/left nipple and complete the splice.

Right/left couplings and nipples also come in handy in remodeling (see p. 202), where new fuel-gas supply piping must be joined to an existing system.

TESTING AND LOCATING LEAKS

Fuel-gas piping is usually tested when the entire system is in place and capped off. You might want to ask your local authority at what pressure the gas lines must be tested. My major code says fuel-gas piping must be tested at 10 psi for 15 minutes or 6 in. of mercury, held for whatever time the inspector asks for. (The mercury test is a lower psi test.) However, many inspectors request a test at the highest pressure the gauge will measure — usually 30 psi.

While the system is pressurized, take a look at your gauge. If the pressure drops, you have a leak somewhere in the system, and you have to check all the pipes for leaks, usually by spraying a mixture of liquid dishwashing detergent and water around the joints in the pipe and fittings. You can make your own mixture (four parts water to one part detergent is good) or purchase a premixed solution at the plumbing supplier. If you have a big leak, you will see giant bubbles that grow to maybe apple size and then pop. If you have a tiny leak, you will see a lot of tiny bubbles slowly growing at the joint. Large or small, if a leak causes your system to lose pressure, it has to be repaired.

There are various ways to pressurize the fuel-gas system. One low-tech method is to use a common bicycle pump, but there are drawbacks. On a long or large-diameter run, it takes a long time to get up to test pressure, and your back gets awfully tired. You really need two people here if you have a leaky system: one to keep the system pumped up while the other looks for leaks.

If you have an air compressor with variable delivery pressure, consider using it for testing. For a one-time filling of the system with test air, this works fine because you can turn off the compressor when the desired gauge pressure is reached. If the system leaks, your compressor needs to be regulated to provide a set low pressure (10 psi to 30 psi). Many compressors for air tools will not self-regulate in this low pressure range. Never test a system at the normal pressure of most air-tool compressors (100 psi or 125 psi) because your system might not hold at such a high pressure and thereafter will fail even at lower pressures.

An airbrush compressor hooked up to a bicycle-pump hose can be used to pressurize a fuel-gas system for testing. An air gauge measures the pressure.

I use an artist's airbrush compressor for pumping up my systems. My compressor, manufactured by Paasche, is small, highly portable and quiet. I adapted it to a standard hose on a bicycle pump, using a simple threaded fitting from the hardware store. In this way I can turn the pump on and save my back. This pump has a relief valve that actuates at a relatively low 30 psi. I can leave the pump running for as long as I need to while looking for leaks in a system with my spray bottle of soapy water.

GAUGES

Fuel-gas testing requires a gauge to measure the pressure in the pipes. There are two types of gauges you can use: air gauges and mercury gauges. An air gas-test gauge has 3/4-in. female pipe threads for threading onto a nipple and a tire valve stem for attaching the hose from a tire pump. Many new air gauges leak, so when you buy one, save your receipt in case you have to exchange it.

Some communities require a mercury test gauge instead of the air test gauge for pressure-testing fuel-gas piping. A mercury gauge costs considerably more than an air gauge, and people steal them off piping because of the mercury's scrap value. The Pasco #3024 12-in. column gauge is a good one for testing both LP gas and natural gas and has instructions that are easy to understand.

I sometimes need to test a section of pipe before the entire system is completed due to time or structural limitations for working in a particular area. I cap off any branch lines and thread my air test gauge to the main line. If I have any valves in the system, they are full-port ball valves, so I need not worry about them holding air-test pressure.

CHAPTER
8

VENTING GAS WATER HEATERS AND FURNACES

If you're using either natural gas or LP gas, you need to convey the combustion gases safely out of the house. This chapter will describe typical venting options for standard residential gas water heaters as well as the installation and venting of stand-up wall furnaces. Each community has its own regulations regarding the venting of gas-fired appliances. As always, it's a good idea to consult your local authority before purchasing any materials or making design decisions that would be difficult to modify later.

Gas appliances are usually vented individually, but two or more appliances can be connected to a common gravity vent if certain conditions are met. If you wish to incorporate any combination vent system in your structure, you should check with your local mechanical inspections department for size and design requirements in effect for your locale. Most codes require the appliances to be on the same floor in the building. The appliances' vent-connector inlets must not be directly opposite each other where they connect to the common vent. The diameter of the common vent must be at least the area of the largest vent connector (individual pipe from each appliance to common vent) plus 50% of the areas of additional connectors. Codes also want each individual vent

connector to have the greatest possible rise (upward slope) that you can provide with the headroom available between the appliance and the point of connection to the common vent.

As a rule, plumbers plumb only for gas venting. Powered hood systems for kitchen ranges are usually installed by the general contractor or the electrician.

MATERIALS

To vent gas appliances, three types of piping are used: triple wall (chimney), double wall (Type B) and single wall (conduction piping). Triple-wall pipe has an inner liner of stainless steel and two outer liners of galvanized steel or stainless steel. Water heaters that are fired by fuel oil, coal or wood are usually vented with triple-wall chimney pipe (consult your local authority for details). Double-wall pipe is galvanized steel with an aluminum liner. Water heaters and gas furnaces are generally vented with Type B pipe, a system approved for both natural gas and LP gas. Single-wall conduction pipe is the piping used to go from the draft hood of the appliance to the beginning of the Type B vent system, either in a wall, outside the building or up through a ceiling.

TRIPLE-WALL PIPE

Triple-wall pipe is used mainly for chimneys, so plumbers rarely work with it. However, it can be used in place of Type B pipe, but it costs considerably more. For residential construction, most triple-wall pipe is used to vent wood, coal and oil-burning appliances. The pipe is assembled it almost the same way as Type B pipe, but inspectors pay particular attention to clearances to combustibles, usually 2 in.

TYPE B PIPE

Type B pipe comes in two shapes, round and oval, and in lengths of 6 in., 12 in., 18 in., 24 in., 36 in., 48 in. and 60 in. In residential construction, for venting water heaters with 20-gal. to 80-gal. holding tanks, the most common pipe diameters are 3 in. and 4 in. (On oval pipe, "diameter" refers to the long axis.) Most codes want Type B pipes located at least 1 in. away from combustible materials. I like to see a bigger safety margin, so when structural conditions permit, I maintain a 2-in. clearance.

CONDUCTION PIPING

Conduction piping is made in three materials: galvanized steel, copper and aluminum. You'll have to ask a wholesale plumbing supplier or the local authority which is allowed in your area. In the San Francisco Bay area, where I work, I can use galvanized steel and copper, but copper conduction pipe is so expensive that it is hardly ever used. I am not allowed to use aluminum, but in the rugged mountain areas of California aluminum conduction piping is quite common. These areas have no natural gas service and LP gas is the only gas available.

My major code wants galvanized-steel conduction pipe up to 5 in. in diameter to be a minimum of 28 gauge, and pipe up to 9 in. in diameter to be a minimum of 26 gauge. Most of the conduction and Type B pipe used in residential plumbing is from 3 in. to 5 in. in diameter. Further, the pipe must be no longer than 75% of the length of the double-wall portion of the vent system.

Conduction pipe has be be screwed or riveted together. Because it gets hot enough to start a fire if it touches combustible materials, most codes will not allow conduction pipe to originate in or pass through an unoccupied attic or concealed space. Most codes want a minimum 6-in. clearance between conduction pipe and combustible surfaces. Because of the fire hazard, conduction piping should be well supported. It's not an area the inspector is likely to ignore.

You can purchase conduction pipe in two forms: preformed and buttonbead. Each has its advantages, and I carry both types on my truck.

Type B vent pipe and fittings are used to vent gas water heaters and furnaces.

PREFORMED CONDUCTION PIPE Preformed pipe comes in 10-ft. long sections. I cut it with my Lenox hacksaw with a fine-tooth blade in the "draw" position (on the back stroke). Conduction pipe at 28 gauge is so thin that if you try to cut it in forward strokes you will squash the pipe and maybe even tear it. Conduction pipe has to be cut as quickly as possible once the blade goes through the top surface of the pipe. If you try to cut the pipe any slower, the blade will bind and the pipe will be squashed.

BUTTONBEAD Buttonbead is flat sheet piping whose edges have been specially formed into male and female configurations. To assemble the pipe you roll the flat sheet into a circle, as shown in the photo below, and then get the male edge to interlock with the female edge. When the edges mate, you hear a loud click. The advantage of buttonbead is that you can cut the pipe to length when it is still a flat sheet, using tin snips, which is a lot quieter and easier than using a hacksaw on preformed round pipe, as described above. However, the longest length of buttonbead commonly available is 5 ft., and it's difficult for a person working alone to roll a sheet this size into a pipe. However, buttonbead is handy for cutting short nipples up to about 1 ft. long.

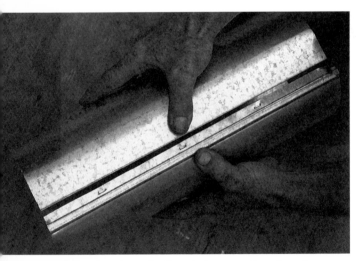

Buttonbead conduction piping snaps together along its edges.

VENTING A GAS-FIRED WATER HEATER

There are three ways to vent residential water heaters: through the ceiling, through an interior wall and up an exterior wall (along the outside of the building). Vents through the ceiling and up an exterior wall use round piping, with just a few differences in support parts. Vents through an interior wall use oval pipe and fittings. Through-the-ceiling vents are used primarily in single-story construction, where the vent can pass through an attic space above the ceiling before going through the roof. That's not to say that you can't have a water heater on the top floor of a multi-story building where you would go through the ceiling and roof. Even in multi-story construction with a water heater on the first level, a properly sized vent-pipe chase that starts at the the water heater location and goes up several stories will still allow the water heater to be vented through the lowest ceiling.

Exterior-wall vent systems can take more time to install than through-the-ceiling vents, because you might have to punch holes through grouted cement block or thick stucco walls and work off ladders, lifts or scaffolding that take time to set and secure. Venting up an exterior wall is usually a retrofit system used in remodel and addition work. It does not win any beauty contests. Where I live, there are lots of two-story additions, and up an exterior wall is often the only path. Even when other routes are possible, up the exterior is usually the cheapest way to get a vent up two stories because it involves only the plumber's labors, not work by the other trades as well.

Venting through an interior wall is almost always done on new construction before the stud walls are covered with drywall. Because you are required to use oval pipe and fittings, there are some limitations in how you assemble the pipe. A round elbow can be swiveled 360° to whatever position is desired, but oval fittings can go only widthwise in a stud bay. The only good thing that I can say about oval pipe is that you can legally get a vent in a 2x4 interior wall, even though the clearance between the outside of the pipe and the inside of the drywall is only about ⅜ in. This does not meet the 1-in. minimum recommended by the manufacturer or technically meet my major code, but it is allowed where I live. I prefer not to vent through interior walls, even though it is allowed. When venting high-consumption Btu burners, the walls can get very hot, so I prefer to play it safe with an exterior-wall vent.

VENTING THROUGH THE CEILING

In order to vent a water heater through the ceiling, the appliance must be located in an area where there is no structural obstacle to a vertical vent path and only one ceiling to go through to reach the roof (or below a chase that extends up to the roof). Wherever the water heater is located, begin by setting up a Smitty pan, if you have one (see p. 153), and position it so the water heater will have a minimum 2-in. clearance from any walls. Rotate the pan so that you can run the drain line once the heater installation has been completed. If you run the drain now and end up having to move the heater several inches to accommodate a vent change, you'll have to cut the drain and redo it. Now, with a few helpers, pick up the heater and set it down in the pan.

Using a plumb bob suspended from the ceiling, find the center of the heater's draft hood and mark the ceiling. Place a tarpaulin over the heater to keep dust and debris off it and out of the flue, then cut a hole in the ceiling about 5 in. or 6 in. in diameter, centered on the mark. Now, from the attic space use your plumb bob again to find the center of the water-heater draft hood. Mark the roof sheathing at this point. If you have a chase and it's a tall one, you might need someone standing right next to the heater to tell you when the bob is centered over the heater's draft hood. Or take binoculars with you.

Using your plumb-bob mark on the underside of the roof sheathing as the center, scribe a circle at least 2 in. in diameter larger than the outside diameter of the double-wall Type B vent you are installing. If the roof is over plywood, it is easier to saw the hole and you won't need additional backing. If the roof is sheathed with shingles over 1x boards that are separated by several inches, you might add some perpendicular backing (same-dimension material), tying together the pieces of sheathing that will be severed when you cut the hole. You might also add backing farther out from the hole, underneath the edges of the roof flashing, to provide something more solid than shingles between the sheathing boards in which to screw down the flashing from above.

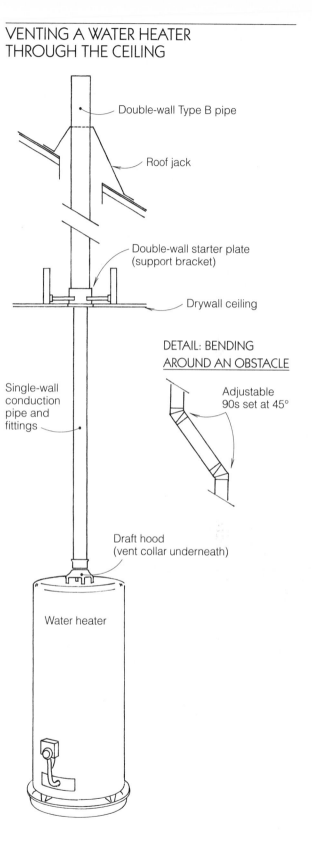

VENTING A WATER HEATER THROUGH THE CEILING

Double-wall Type B pipe

Roof jack

Double-wall starter plate (support bracket)

Drywall ceiling

DETAIL: BENDING AROUND AN OBSTACLE

Adjustable 90s set at 45°

Single-wall conduction pipe and fittings

Draft hood (vent collar underneath)

Water heater

VENT FLASHING ON THE ROOF The through-the-ceiling venting system will use a single-wall conduction pipe to go from the top of the heater's draft hood up to the beginning of the double-wall Type B vent, which will start at the drywall ceiling or opening of the chase. But before any pipe is joined, the roof jack, as it is called in some areas, has to be installed. The roof flashing will act as a target for the pipe run.

Sometimes the roof jack may be installed by the general contractor or roofing contractor when the roofing system is installed; other times it is left to the plumber to do after the roofing is in place. If the roof is already completed, the task is a bit more laborious.

With fiberglass or asphalt shingles, slide the leading edge of the jack up under the shingles until the hole in the jack is centered over the hole in the roof, with the long slope of the jack's cone on the down side. You can use the plumb bob from above the roof to check for this alignment. When the jack is centered, trace along the edge of the exposed sides and bottom of the jack on the shingles to mark its position (I use a soapstone, but a felt-tip marker would also work).

After sliding the jack back out, punch four holes per side along the edges of the jack to receive the screws, using a hand punch (see the photo on p. 186) or a cordless drill with an ⅛-in. bit or bit-tip screws. Then butter the underside with a generous layer of black roof mastic, reposition the jack and screw it in place with zinc-coated drywall screws that are long enough to bite into the roof framing. (Stainless-steel panhead sheet-metal screws work even better.) On wood-shingled roofs, it's a good idea to predrill the shingles to prevent splitting before driving the screws. With asphalt or fiberglass shingles or with composition or membrane surfaces, predrilling is unnecessary.

If the roof pitch is steeper than 4-in-12 and not on the weather side (the predominant direction from which rain, sleet and snow come), I do not drill and screw across the bottom edge of the jack. On shallower pitches on the weather side, I do screw down the bottom edge. But in all cases, I diligently cover each screw head with a layer of mastic. If the roof jack is highly visible, I paint the mastic with aluminum paint.

PIPE INSTALLATION AND SUPPORT Through-the-ceiling vent pipes need to be supported, and if they pass through a chase they also need a fire stop. Fire stops and support brackets are supplied by each manufacturer of double-wall Type B gas vents. The designs differ, so you need to refer to the product installation instructions for information on how to install blocking or framing to work in conjunction with these components. Chases need fire stops and starter plates at each story. If the vent passes through only one ceiling and one attic, you will need only one starter plate and fire stop at the ceiling level, but these will require proper framing and blocking.

With the fire-stop/starter-plate assembly and the roof jack installed, you now have a target for the piping. You can bridge the two components with Type B vent pipe, starting from the bottom. Any support brackets will nest the pipe in a centered position within them.

Sometimes the vent can not rise straight vertically through an attic because of a ceiling joist, diagonal brace or masonry chimney, so you have to use 45° adjustable elbows. When you have some vertical piping in place, then add a change-of-direction fitting and once more proceed upward, the fitting may swivel askew. Neither the piping manufacturer nor your local authority wants you to use sheet-metal screws in the joints to prevent the fitting's parts from rotating. Screws can cause the adjustable joints to separate, and they also pose a puncture danger to the inner metal liner of the pipe.

If movement of the fitting is a problem, many local authorities recommend soldering the fittings in the desired position. However, soldering can be time-consuming if the house has a lot of elbows, so sometimes I just use the solid aluminum tape made by the Nashua Corporation to tape these joints in the desired position. I sometimes have to use a rag soaked in vinegar to remove any oil film on the outside of the pipe and/or fitting prior to applying the tape.

Piping that leaves a vertical path can be supported by plumber's tape. I wrap tape around the pipe, install a nut and bolt through two matching holes in the tape and then extend the tape to purlins or to rafters, using my cordless driver to secure it. Two or three strands of tape extended in different directions will support the pipe in a stable manner.

A starter plate supports a section of Type B vent pipe as it rises toward the roof. (Photo by Bill Dane)

Type B vent pipe and fittings have male and female ends, and when the two are properly joined, you should hear a snap or a click, depending upon the manufacturer. The male end of the pipe should always point up so that moisture can be shed without working its way between the two layers.

Adjustable elbows can change the direction of the vent to avoid obstacles.

CONNECTIONS AT THE ROOF   At the top of a vent pipe, where the run ends, the pipe sticks out of the roof. Building codes specify where on the roof the pipe should terminate and how high above the roof the pipe should rise. My major code wants Type B pipe brought through the roof at least 4 ft. from any portion of the building that extends upward more than 45° from horizontal, such as gables, valleys of steep roofs and widow's walks. Often the architectural design of your structure makes it difficult to comply with this requirement. If you can't figure out a way to do so, you will have to get your local authority's approval to deviate from code.

My code calls for a minimum termination height of 1 ft. off the roof when no special conditions exist. The manufacturers of Type B pipe have their own recommendations of termination heights, which are usually higher, and local authorities generally go by manufacturers' recommendations. When you purchase materials, you can request free installation guidelines, which will include the manufacturer's recommended termination heights.

The roof jack (left) slips under shingles to support the last piece of vent pipe; silicone sealant keeps out the rain. A storm collar (right) further protects the joint. (Photos by Bill Dane)

With the pipe bridging the lower starter plate and fire stop and penetrating one pipe length (any length) out the top of the roof jack, it is time to add the storm collar and the cap. I seal the joint of the roof jack and pipe with silicone sealant before bringing the storm collar all the way down until it almost touches the siliconed joint, as shown in the photos above. Then I use silicone sealant again to bridge the gap between the collar and the pipe. If more sections of pipe are required to bring the vent to the required height above the roof, I install them next. Finally I install a cap on the top end of the uppermost pipe. If the pipe protrudes above the roof 5 ft. or more, you should make sure that it is securely braced. Triangulated guy rods or wires provide good stability, as do two rigid rods.

INSTALLING THE VENT CONNECTOR   With the major pipe run complete, the vent connector can be installed between the water heater and the Type B vent. The draft hood for the water heater will have a short, vertical top edge, probably about ¼ in. to ⁵⁄₁₆ in. high. Many hoods will already have a hole punched through this edge. The hole is for receiving a sheet-metal screw to secure the start of the vent connector. If there is no hole, just drill one through both the vent connector and the top of the hood and install a screw. I use Malco's "bit tip" sheet-metal screws that have a drill point on the end; they drill their own hole, so I don't need the drill bit. I put a magnetized ¼-in. hex nut driver bit in the drill chuck to drive these screws.

If your Type B pipe is larger in diameter than the water-heater's vent-connector pipe, you will need to install a single-wall adapter bushing into the Type B pipe before cutting a cut-to-length piece of conduction pipe for the vent connector. The adapter bushing slips into the open female end of the Type B pipe.

VENTING THROUGH AN INTERIOR WALL

Double-wall Type B oval vent pipe is used for water-heater (and furnace) vents up through a 2x4 interior wall. Many water heaters are legally vented with 3-in. dia. pipe, but oval pipe comes only in 4-in., 5-in. and 6-in. diameters. Most residential holding-tank water heaters up to 50 gal. in capacity can be vented with 4-in. Type B oval pipe. For any other type of heater, it's best to confer with your local authority and the representative or agent of the heating device's manufacturer regarding the size and type of vent pipe to use.

There are two possible scenarios involving the starting run using oval pipe and fittings. In the first, the water heater stands on a lower floor with a ceiling above it, and the wall in which the vent is to pass spans the ceiling. In the second, the water heater stands close to a wall, and an upper-story wall is directly over the lower wall. Once the vent enters the vent wall, however, there is no difference between the two scenarios.

VENTING THROUGH A CEILING AND A VENT WALL If a ceiling and the vent wall are above the water heater, you begin by locating and cutting the hole in the ceiling above the water heater, as described on p. 179. Then install either a starter plate (Metalbestos) or a bucket support (Amer-Vent), depending upon which product line you are using. Below the hole, also install a bushing for single-wall pipe, to which the vent connector will be attached (see the discussion above). You then place a round-to-oval adapter on the starter plate or bucket. With the wall's lower plate cut away on the floor above, you install the fire stop (see the detail in the drawing at right) and start your run of oval pipe off the round-to-oval adapter.

In a wall going straight up, you should be using the longest available sections of oval pipe, usually 5 ft. When you go through the upper plate on the vent

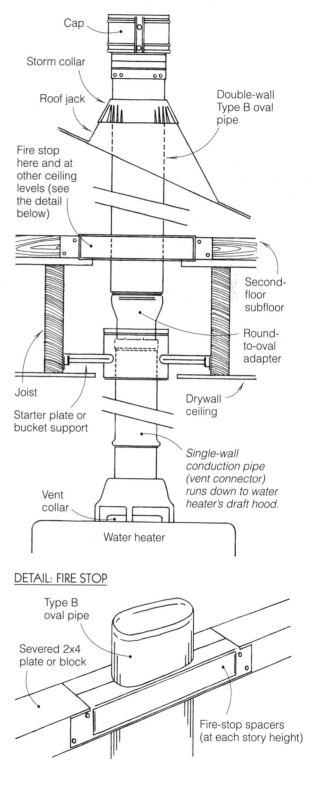

VENTING A WATER HEATER THROUGH A CEILING AND A VENT WALL

Cap

Storm collar

Roof jack

Double-wall Type B oval pipe

Fire stop here and at other ceiling levels (see the detail below)

Second-floor subfloor

Round-to-oval adapter

Joist

Drywall ceiling

Starter plate or bucket support

Single-wall conduction pipe (vent connector) runs down to water heater's draft hood.

Vent collar

Water heater

DETAIL: FIRE STOP

Type B oval pipe

Severed 2x4 plate or block

Fire-stop spacers (at each story height)

wall, you install another fire stop. Once in the attic you have the option of going back to round pipe by using an oval-to-round adapter, or you can keep going out the roof with oval pipe and use an oval roof jack, storm collar and cap. (Oval pipe often costs more than round, so if you need a lot of material you might want to return to round.) You use a plumb bob to mark the underside of the roof for cutting the passage, just as for round pipe. There are no oval 90° elbows, only fixed 45° elbows. In an attic space, to go around an obstacle, you use the 45° fittings and support your offset with plumber's tape, just as with round pipe.

VENTING THROUGH A VENT WALL   If the vent wall is behind the water heater, instead of going off the heater's draft hood up to a ceiling, you use a 90° elbow off the draft hood in round conduction pipe and go to a wall and greet a round tee branch on an oval tee; from this point up the entire system is oval pipe and fittings, as shown in the drawing at right. A wall plate or tee collar will conceal the hole in the wall where the tee passes through. As shown in the drawing at right, a component called a tee support is nailed to a horizontal 2x fire block. The Amer-Vent tee support has two punched buttons on its walls that snap into the grooved bottom of the oval tee and secure it to the support. The fire-block height is so set that the tee, resting on the support, has a branch protruding out of the finish wall at a height that provides ample elevation above the heater to use standard Type B or single-wall pipe and fittings to reach over and down onto the water heater's draft hood. For fire safety, though, you want to keep the horizontal vent connector as far below the drywalled ceiling as practical.

Vent runs can't always go straight up. If your pipe needs to change direction to the left or right (say, to go from one stud bay to the next), you use a flat elbow. For offsets in front of or behind the run (say, to avoid another pipe), you use a standard elbow.

VENTING A WATER HEATER THROUGH A VENT WALL

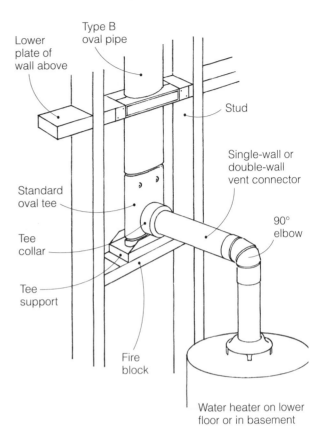

Lower plate of wall above

Type B oval pipe

Stud

Standard oval tee

Single-wall or double-wall vent connector

Tee collar

90° elbow

Tee support

Fire block

Water heater on lower floor or in basement

VENTING THROUGH AN EXTERIOR WALL

The third option for a water-heater vent is up the outside of a building, and the rules about clearances to combustible materials (see p. 177) still apply. Try to find an exterior vent wall that is not visible from the street or from within the building; no windows should be located in the proposed path of the vent. Exterior vents are often run on a hidden side of the building or on the far side of an exterior masonry chimney or other offset in the exterior wall. Sometimes, though, structural considerations leave only one possible choice for the exterior vent wall.

FROM WATER HEATER TO THIMBLE   With the draft hood fastened to the top of the water heater, begin by fitting a single-wall 90° elbow onto the top of the hood, as shown in the drawing at right. This is usually an interference fit, which means you might have to push firmly on the elbow to get it to slide over the top ridge on the draft hood. Usually you will have enough friction between the two parts to let go of the elbow. A 40-gal. or 50-gal. holding-tank water heater will have a 3-in. or 4-in. opening on top of the draft hood. Use the appropriately sized single-wall 90° elbow. Using a level, mark the wall at a height matching the center of the horizontal opening of the elbow. Then raise the mark ¼ in. for each foot of distance to the wall. Because this is a gravity vent system, it will operate more efficiently if the vent connector slopes slightly upward instead of being perfectly horizontal.

Where the vent connector meets the exterior wall, you have to cut a hole and install a thimble. The thimble is a two-piece telescoping spacer that holds Type B pipe passing through the wall away from combustible materials. In my area, where mild weather prevails, most exterior walls are 2x4. Most factory-made residential thimbles sold here generally fit 2x4 to 2x6 walls. If you live in a cold climate, the local sheet-metal suppliers will probably stock deeper thimbles, which you will need if your wall is thicker. Some thimble sleeves are smooth and merely push in and pull apart; others are spirally corrugated and thread in and out. Whichever types you find, you should handle them gingerly; they are only spot-welded and break apart all too easily.

If the wall you are going through is drywalled, probe around your mark on the wall with an electronic stud finder or a tiny drill bit in your drill to see if there is an unobstructed void in the wall that is large enough to accept the thimble. Always try to avoid cutting a stud when installing the thimble. Sometimes you can move forward or backward on the wall far enough to get a properly sized hole and then use an additional adjustable single-wall 45° elbow in conjunction with a vertical single-wall 90° elbow to swing forward or backward to reach the off-center thimble, meeting it in a true perpendicular angle. If you use this ploy to avoid cutting a stud, remember that single-wall fittings and pipe must be

VENTING A WATER HEATER THROUGH AN EXTERIOR WALL

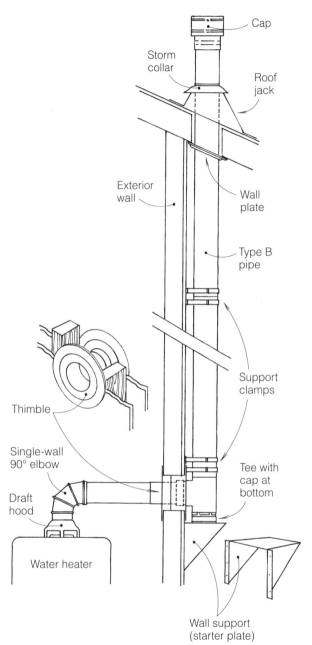

Cap

Storm collar

Roof jack

Exterior wall

Wall plate

Type B pipe

Support clamps

Thimble

Single-wall 90° elbow

Draft hood

Tee with cap at bottom

Water heater

Wall support (starter plate)

A hand punch makes holes in the thimble flange.

kept at least 6 in. away from the walls or other combustibles. If you can't meet this last requirement, you can use the same technique using Type B double-wall fittings, which in most communities will require only a minimum 1-in. air-space clearance to combustibles. If that doesn't work, you'll have to open up the wall and install some new framing.

Once the hole is laid out and cut, you can install the thimble. I usually do this job alone, but it's a lot easier if you have two people, one inside the building and one outside. The larger-diameter sleeve goes in from the exterior side, and the smaller-diameter sleeve fits inside it from inside the building. Before you begin, punch three holes around the edge of each half with a hand punch (see the photo above). Then, using stainless-steel panhead sheet-metal screws, screw the larger-diameter sleeved half of the thimble in the hole on the outside wall and from the inside of the building maneuver the smaller half into the half that is already mounted. When the smaller-diameter sleeve is in place, screw its flange to the wall.

INSTALLING THE TEE AND STARTER PLATE Now go back outside and install the wall support, which is also called the starter plate. The hole in the thimble is sized for double-wall pipe. The first step is to get the tee temporarily assembled so that you can determine the height at which to install the wall support. From inside, on a 2x4 wall, nose a 6-in. long section of pipe

(male end first) into the hole in the thimble and carefully work it in until it starts coming out of the thimble on the outside. On a 2x6 wall, you'll probably need to use a 12-in. long pipe section. Push on it until you are up to the recessed ring that runs around the female end. The pipe is rarely perfectly round but the hole in the thimble almost always is. Work slowly because the pipe may be so out of round that it won't slide easily through the hole, and you can break the flanges off the thimble sleeves trying to work the pipe all the way through both openings.

At this point, go outside and see if you have enough male pipe sticking out of the thimble to penetrate partway into the female branch of a tee. On the 12-in. long pipe piece you surely will. This is where working with a helper really saves time. You want to get the tee's branch pushed onto the male pipe far enough to determine that the branch is at a true level position without joining the two parts all the way and activating the locking mechanism (if that happens, getting them apart again can be difficult). Now back outside. With the tee still hanging there in midair, gently shove it inward until the branch of the tee is all but about 1 in. from being totally inserted in the hole. Now lift up the starter plate (wall support) until it contacts the bottom barrel (or trunk) of the tee. I have seen these supports mounted in two positions: legs down, flat surface on top and legs up, flat surface on the bottom. The manufacturer's literature shows the support mounted legs down. But I like the legs-up position because it affords some impact protection for the tee.

Now mount the starter plate on the wall in the position you decide on. If the exterior wall is stucco, you might have to use masonry bits and expansion anchors. When the starter plate is secured, adjust the depth of the tee into the thimble until the legs of the support clamp contact the exterior wall of the building. Leave the tee in this position.

ROOF JACK AND PIPE ASSEMBLY For a perfectly straight vertical run, get your ladder or scaffolding in place and drop a plumb bob from the underside of the eave or soffit to the center of the top barrel of the vent tee on the starter plate. Mark the spot on the eave or soffit where a hole will be cut for the pipe to pass through. If the soffit is 1x lumber or plywood, I use a sharp hole saw and bore about a 1½-in. dia.

hole. Then using a "sacrifice" auger bit in an extension, I continue up through the asphalt, fiberglass or built-up composition roofing. For wood shakes or shingles, I bore this top hole with the same sharp hole saw in the extension.

You will need to enlarge the hole in the eave or soffit to allow for air clearance around the pipe. For cutting through asphalt, fiberglass or built-up composition roofing, a new sharp 6-in. rough-in blade in a reciprocating saw works well; frequent wipes with a kerosene-soaked rag will keep the blade clean. For cutting through wood shake and shingle roofs, I like to use a jigsaw, for better control. With the hole cut through the roof, you can temporarily screw the roof jack down (see p. 180) without any roof mastic applied to the bottom.

You are now ready to install the pipe sections. Begin by extending your tape down to the tee for a measurement. After consulting the table in your product's installation and assembly instructions for vent height above the roof as a factor of roof pitch, divide the entire distance up into individual pipe lengths. As with other pipe runs, the fewer the joints, the better. When the pipe is assembled, the joints overlap, so in figuring length you should de-duct 1 in. for each pipe joint. If by using the available lengths you cannot arrive at the exact measurement, you should use the next longest length, which will be longer than you actually need. If you do not go to the next longest pipe length, you won't be installing the vent to the manufacturer's recommended termination height. Usually just a few inches too short doesn't make much of a difference. But if you want your material warranty to be honored, follow the manufacturer's recommendations to the letter.

Now it's time to put the first length of pipe on top of the tee. Set the female end of the pipe onto the discharge end of the tee (which is male) until you hear a snap or click, or until you see that the two are fully mated. As you assemble the pipe sections, you'll need to support them at regular intervals with support clamps. These clamps are split in the center, with a nut and bolt to draw them tight to the pipe. I position the clamps with the nut and bolt on the inside, near the wall. On the clamp are riveted two legs, which pivot on their rivet.

The water-heater vent passes through the thimble to the exterior, where it meets a tee supported by a wedge-shaped starter plate. An adjustable 45° elbow angles the vent around a window.

Clamps above and below each joint hold the vent pipe securely to the wall.

Most manufacturers of double-wall pipe suggest that the pipe be supported to the structure at least every 6 ft. Your local code might be more stringent. The clamps are so inexpensive that it doesn't pay not to do a really good job anchoring the piping down. As shown in the photo above, I use one clamp above and below each joint, beginning with the tee, regardless of how short the length is. Since 5 ft. is usually the maximum length of any Type B vent, with a clamp at each joint you will have a well-supported system that your local authority will approve of.

Slide the clamps down the pipe sections as you snap the pipe together, starting from the tee and working up, with the seam on the back side of the pipe (close to the building). Before you go through the soffit or eave with the pipe, slip a flat wall plate down over the pipe with the finish side down. You can use a hand punch to put three holes near the outside edge.

When snapped together, the pipe will not automatically be held perfectly true. The pipe can still move slightly, even when fully mated to another piece. When all the pipe sections have been assembled, start at the bottom with a torpedo level and true up each pipe section and screw each clamp to the structure. On stucco exteriors this work takes longer because you have to mark the holes, drill for anchors and finally screw down the clamps.

When the run trued up, slide the wall plate up and screw it to the soffit or eave to hide the rough hole. In the case of an eave without a soffit, sometimes the plate won't angle enough to match the pitch of the roof unless you first enlarge its center using tin snips. At this point you can lift the roof jack and butter up its underside with roof mastic and screw it down, as described on p. 180. If you live where snowfall is heavy, your jack's cone should be taller than the standard jack's. Before adding any more sections on the pipe section protruding out the top of the roof jack, shove the storm collar down past the male end. Then add whatever additional pieces of pipe are needed to reach the proper termination height of the vent.

I have found it a good practice to use some solvent or vinegar and remove any grease or oil on the jack and the pipe section as it passes up out of the jack. I also degrease the top opening of the jack itself. Then I apply a heavy bead of silicone sealand around, in and on the joint, slide the storm collar down to within an inch or so of the silicone, and put another bead around the joint of storm collar and the pipe, as explained on p. 182. The last step is to snap on the cap.

Sometimes you have to trim the hole in the soffit or roof a tad to get the pipe run perfectly plumb, but that is not a problem — the wall plate and the roof jack have plenty of square area to cover even the modified holes. If you are going through a narrow eave or soffit, you might have to trim off part of the flat rim of the roof jack so that it does not overhang the gutter or edge of the roof. Or you may have to trim off part of the wall plate to get close enough to the wall to center it around the rough hole.

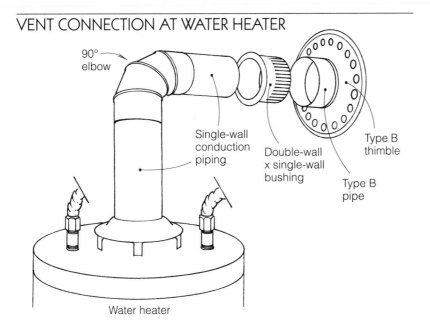

90° elbow

Single-wall conduction piping

Double-wall x single-wall bushing

Type B thimble

Type B pipe

Water heater

CONNECTING TO THE WATER HEATER Back inside the structure, at the water heater, there is a short section of Type B pipe that is yet not fully inserted into the tee. The drawing above shows a cross section of the wall with the tee held in place at the thimble and the wall support just below it. Part of the pipe penetrates into the tee, and the female end protrudes out of the interior side of the thimble. In 2x4 walls, this pipe would be 6 in. long; in 2x6 walls, this pipe would probably have to be 12 in. long.

Usually the protruding pipe is short enough to leave plenty of room to add a double-wall by single-wall bushing and then use a single-wall adjustable 90° elbow to drop down onto the draft hood, maybe with the addition of two short pieces of single-wall pipe. However, sometimes you might find yourself with too much pipe sticking out of the thimble to add the 90° elbow and be on target with a small water heater's draft hood. When that happens, you have no choice but to move the water heater farther away from the wall, if you want your installation to follow code.

If the length of the pipe through the thimble is not a big factor, you could even continue the run with double-wall pipe, either another 6-in. long section or maybe even a 12-in. long section, and then add a double-wall 90° elbow and go down onto the draft hood with another piece of double-wall pipe.

Double-wall pipe is acceptable in this application, and it lasts longer than single-wall pipe when used as a vent connector.

INSTALLING AND VENTING A STAND-UP WALL FURNACE

In mild climates, houses don't need full-fledged furnaces, but on chilly evenings a little heat can make things more comfortable. In my locale, the stand-up wall furnace is a popular alternative to the large furnaces common in cold climates.

Stand-up wall furnaces are small gas-fired heaters that are installed in a wall to heat one or more rooms. Because they sit inside the wall, they are usually installed at the same time as the rough plumbing. (Other heating appliances, such as direct-vent heaters, are hooked up later, after the walls are closed in. Installation procedures for these and other gas appliances are described in *Installing and Repairing Plumbing Fixtures,* the companion volume to this book.)

As usual, it pays to select the model before you run your gas line. Various brands have their burner and control valves in different locations, and their rough-in instructions or template will suggest the proper gas line stub-out location for that particular appliance. If you run the gas line first without considering which brand of appliance that you will be using, the gas shut-off valve that you install may not be operable with the furnace installed, or the furnace may not even fit in the rough opening because the gas valve is in the way.

Stand-up wall furnaces are designed to fit into a standard 16-in. stud bay and are partially set into the wall. Exactly how far into the wall you set the furnace will depend on the model and the design of the trim package, or cabinet, which consists of the outer, baked-enamel sheet-metal shell with a grill. The shell is held to the furnace by long sheet-metal screws at the top and by a screw or two at the bottom. If you set the furnace into or out of the wall too far, the trim will not fit properly, so pay particular attention to the manufacturer's installation instructions when you measure for depth.

Most furnace models are screwed or nailed in the wall through holes punched on each side. The two (in-the-wall) sides of the furnace will actually be supporting members, since the bottom of the sides have legs that rest right on the finish floor or lower plate.

The average stand-up residential furnace is rated from about 20,000 Btu to 50,000 Btu heating capacity. Some of the upper-end capacity heaters are rated 50,000 Btu to 65,000 Btu and may be two-sided; that is, they will have a heating surface opening into two rooms, one on each side of the wall.

One cost-effective way to increase the efficiency of the furnace is to install a blower. A lower Btu furnace with a blower will do a better job of heating a space than a higher-capacity furnace without a blower. If you want a blower, you must settle on the brand before you drywall so that the electrician can string wires to the proper spot.

Residential stand-up wall furnaces under 55,000 Btus can be vented through an interior wall with 4-in. Type B oval pipe. In addition to the usual fitting, you'll need a Type BW hold-down plate, Type BW plate spacers and fire-stop spacers (the designation BW refers to Type B piping, especially for walls).

As shown in the drawing at right, the hold-down plate is secured to the furnace and anchors the starting section of pipe (fire-stop spacers are used at subsequent levels). This sheet-metal part does not come with the stand-up wall furnace, but must be bought

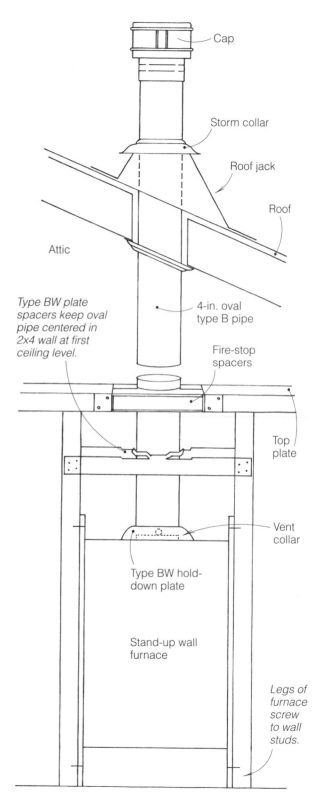

VENTING A STAND-UP WALL FURNACE

Cap

Storm collar

Roof jack

Roof

Attic

Type BW plate spacers keep oval pipe centered in 2x4 wall at first ceiling level.

4-in. oval type B pipe

Fire-stop spacers

Top plate

Vent collar

Type BW hold-down plate

Stand-up wall furnace

Legs of furnace screw to wall studs.

separately. It attaches to the top of the furnace with two sheet-metal screws. The Amer-Vent BW hold-down plate has two buttons, one on each vertical side, that snap into the bottom groove of the vent pipe and secure the pipe in a centered position over the furnace's vent collar. Your inspector will make sure that you have not tried cutting corners by omitting this somewhat flimsy but crucial part.

The spacers center the pipe in the wall at the first ceiling level above the furnace. They do not support the weight of the pipe, which rests on the hold-down plate and is held up by the furnace itself.

A standard framed 2x4 wall with Type B oval pipe running through it does not have 1-in. clearance to the drywall; it is more like ½-in. to ⅜-in. clearance. I can vent furnaces rated below 30,000 Btus and sleep at night, but a 65,000-Btu furnace vented with oval pipe in a 2x4 wall and operated at capacity can get the drywall very hot. If you want to vent a high-capacity stand-up wall furnace through an interior wall, I recommend that you consult with your local mechanical or plumbing inspector before hanging the drywall.

The oval pipe starts on top of the BW hold-down plate. Place the female end down, and use the longest sections of pipe you can to keep the total assembly to the fewest number of pieces. You will usually hear a loud click when the male end of the pipe locks into the female end of the next pipe length. Do not join two pieces until you are ready, for they can be difficult to separate without damaging the joints. Fire stops have to be placed at each story height.

Once the pipe run passes through the plates in the top floor's wall into the attic, you can continue out the roof in oval pipe and use oval roof flashing, storm collar and cap; or you can adapt to round pipe in the attic by using an oval-to-round adapter, exit the roof in round pipe and use its version of roof flashing, storm collar and cap. With round pipe you can use 45° fittings that swivel, which will allow you to move out into space with more headroom to fit the pipe around any obstacles, such as bracing.

If you want to exit the roof in a straight vertical path from the top floor wall's upper plates, you can use a plumb bob to transfer the location of the opening in the upper plate to the underside of the roof. Be sure that your roof opening provides for 2 in. of clearance from the pipe to the edge of the hole. If you al-

ready have pipe strung up this far, make sure to drape something over it to keep wood chips and roof debris from falling down into the pipe.

Consult the furnace manufacturer's literature for the suggested termination height of the pipe for the Btu rating of the furnace. When the pipe is the proper height, you can install the roof jack if it hasn't already been installed by someone else. Wipe the pipe and jack with vinegar to remove any grease or oil. Seal the joint of pipe to jack with a liberal bead of silicone sealant and then slide down the storm collar to within an inch of the silicone. Add another bead of silicone to the joint of the storm collar and pipe, and then install the cap.

VENTING A DRYER

All dryers need venting to the exterior of the building. That's because the dryer produces lint, some of which escapes the appliance's filter and would otherwise collect inside the house. Besides being unsightly, lint is flammable. Also the hot, moisture-laden exhaust air from the dryer can cause serious mildew on the laundry-room walls and ceiling.

Unlike other vents, which rise, dryer vents slope down. Vents for the dryer should run to the exterior of the building at as steep a slope as possible, so the lint will drop out. The system operates best when smooth, rigid piping is used, such as aluminum or galvanized-steel conduction pipe. Changes of direction can be made with adjustable elbows, as discussed on p. 189.

Don't be tempted to use long lengths of flexible ducting to vent your dryer. Lint clings to its internal ridges and restricts the air flow through it, causing dryer performance to suffer. Some dryer manufacturers will void the machine's warranty if more than a short length of flexible ducting is used.

A dryer vent starts at the vent collar (usually 4 in. in diameter), which on conventional full-sized dryers is usually centered on the back side, near the bottom. The vent passes through a thimble and ends on the exterior at a plastic or aluminum vent hood with a hinged flap that seals off the opening when the dryer is not operating. For specific vent-installation instructions, see the manufacturer's literature that comes with the appliance.

REMODELING

A remodel project may involve building an addition or renovating an existing part of the house, and remodel plumbing may involve rerouting old DWV and supply lines or installing new ones. In order to design the new systems, you will need to analyze the existing plumbing. Where can you tap into the building drain or building sewer for gravity drainage? Where can you get hot and cold fresh water? Will code changes require upgrading any valves or fixtures? If you have a need for fuel gas, where can you hook into that? You will also have to decide if you need an additional water heater and determine if the existing pipes are properly sized for the additional demand. You should have answers to these questions before any building takes place. A lot of great remodeling plans have been scrapped because a drainage plan for the new space could not be worked out without the addition of an expensive sewage ejector.

In remodeling, plumbing plans are usually based on compromise. Most codes will sanction deviations from the rules on remodels that they wouldn't even consider on new construction, but obtaining the inspector's approval may call for all the diplomacy you can muster.

Some local authorities may want detailed drawings of your intended piping system and changes called for in the existing system before you start the work. Producing these drawings can be a real problem if demolition hasn't yet begun and the existing pipes aren't exposed. After making the best guesses that you can on where the supply and drain lines are and how you will get them where you need them, you won't know for sure that the work can be done as planned until the existing walls, ceilings and maybe floors have been demolished. During the demolition, work often has to stop for the severing and capping of lines. An experienced contractor will try to remove as much necessary structural material as possible before cutting any piping, especially if the dwelling is being lived in during the project. With major supply lines passing on to other rooms not affected by the remodel, temporary rerouting is sometimes necessary before work can proceed.

In remodeling, you have to contend with differences in materials and sizes between the existing plumbing and the new, which must meet current code standards. For instance, about 20 years ago a code-sanctioned kitchen-sink drain line in my locale had a minimum inside diameter of 1½ in. Today code calls for a 2-in. drain. Consider what might happen if you are remodeling a kitchen and moving the sink to another location, from which you cannot reach the drain with a continuous waste. Your new drain will have to be 2 in. in diameter (unless otherwise negotiated with the local authority), so at the new location, the sink will need to be replumbed with a minimum 2-in. drain, and maybe a new vent if you are now too far away from the original one. When you are on a first floor over a crawl space, this is usually no big deal, but for slab and second floors, it is a big nuisance and very costly.

Even changes that don't involve relocating fixtures can be complex. Let's say you are replacing a wall-hung lavatory with a cabinet-mounted sink in the same location. However, the present drain is a clogged-up 1¼-in. galvanized-steel run that drains very slowly. On a first-floor over crawl space or basement, you wouldn't think twice about changing it to a 1½-in. or preferably 2-in. line. But on slab and multi-floor installations, the cost of the drain-line replacement can be prohibitive. If there isn't money in the budget for this work, the clients may get angry with the plumber when they discover that the new sink doesn't drain any better than the old one. A similar situation arises if new fixtures are plumbed into undersized, antiquated supply lines.

The National Energy Policy Act of 1992 calls for the installation of water-saving fixtures, such as aerators, water-saving shower heads and low-flush toilets, in new and remodel construction. The act also recommends the use of temperature-compensating valves. These anti-scald devices control the mix of hot and cold water, preventing sudden pressure drops and temperature changes. In my area, all the temperature-compensating valves that are sold in my locale are single-handle models. Two good ones are made by Moen and Grohe (see Sources of Supply on pp. 206-209). For people who wish to retrofit their existing, unregulated single- and two-handle sink faucets and tub and shower valves, Resources Conservation makes after-market temperature-compensating faucet aerators, tub spouts and shower-arm regulators under the Scald Safe name. These adapt to your present valves or may also allow you to remodel with the more practical, two-handle valves (check with your local authority).

The installation of new drain, waste and vent piping, water-supply piping and fuel-gas piping for a remodel follows most of the routines and techniques described elsewhere in this book. The biggest difference, and in my opinion the trickiest part of rough plumbing on a remodel or addition, is connecting the new work to the existing plumbing. Supplying water is usually straightforward because it involves few code considerations, beyond proper sizing and a few connection rules. Installing the larger-diameter piping for DWV is usually more problematic.

DRAIN, WASTE AND VENT

Some of the issues that frequently arise in remodel DWV plumbing are dealing with abandoned vents, laying new branch drains and joining them to the building drain, cutting old iron pipe and splicing dissimilar materials. If the remodel is such that wastes can't be made to flow downhill, then a gravity drainage system is out of the question and a sewage ejector will have to be installed.

VENTS

If your remodel does not involve an addition, but maybe just some wall changes and fixture rearranging, then you may have one or more severed existing vents. These may be in an attic, wall space or ceiling-joist space, but they still run upward and through the roof. If you are doing a remodel on the lower floor of a multi-story structure, the vent and drains in your way may well be serving additional fixtures above you, in which case you must find a path for rerouting them to the building drain. If the vent is also part of an upper-floor vent system, you might be able to reuse it if you can route a vent from the new fixtures back to it, as long as you don't overload the vent with new fixtures. Check the fixture-unit tables on p. 50 to see if you can use existing piping and save yourself the expense of running new pipe.

Sometimes you need to go up another floor with a vent for a new fixture or for two fixtures whose vents have been spliced together (back-vented) if the existing vent's diameter is too small to handle the job legally. Can you still use it? That depends upon the views of the local authority. If running a new vent to size would require major structural work to the building, you may be able to get a variance for a reasonable request. You wouldn't want to vent a toilet, shower and lavatory all on one 1½-in. vent, but you might get away with two lavatories on one 1¼-in. vent that rises less than 45 ft. or one kitchen sink on an old, not very long lavatory vent.

If you have an existing vent on a lower floor that is being remodeled and you need to move the vent but it is also back-vented into a higher floor's fixture vent and you are not going to rerun this lower section of vent for the new plan, you have a problem — piping that is open to the sky but no longer extends down to the building drain. In a rainstorm, water would enter an open vent and run freely into the structure. You may be required to continue that vent to the building drain, or you may be allowed to leave it capped in a wall or ceiling, if you do so permanently to the satisfaction of the local authority.

If the plumbing fixture that is being abandoned is on an exterior wall and the vent runs straight up through the upper plates and out the roof, do not plan on being able to connect to this vent in an attic space to reroute it if the roof pitch is low. Unless the roof rafters are 2x12s or 2x14s, you will rarely have room to crawl, much less get a coupling and a 90° elbow into the space between the plate and the roof. In this situation, I would try to cap the old vent and run a new line away from the exterior walls.

If you want to reuse an existing 1¼-in. galvanized threaded-steel lavatory vent for a new lavatory and the severed vent is in an accessible ceiling-joist space, you can use a 1½-in. copper by 1½-in. plastic or steel Mission coupling to join the 1½-in. copper to the 1¼-in. galvanized steel. (The 1¼-in. steel pipe has the same outside diameter as the 1½-in. DWV copper pipe for the sake of compatibility with band seals.) You sever the pipe with a reciprocating saw and end up with a flush-cut end of steel pipe, to which various types of new DWV material may be attached.

Old, existing vent lines usually pass through steel or lead roof flashings sealed with roofing cement. If you have to cut any pipe in attic spaces, it's a good idea to reseal the flashing after reconnecting the vent. Usually the vibration from the reciprocating saw will crack the seal of old mastic, and the roof flashing will leak the first time it rains. If you need to run new vents through the roof, it might pay to hire a profes-

sional roofer, especially if the roof is tile or slate, which require specialized roof flashings. A specialist could probably remove the existing material, install the new jack and replace any roofing materials with less breakage and in a fraction of the time it would take you.

DRAINS

On a house with good access from below (basement or crawl space), you can usually cut into existing lines fairly easily. But it's not always this simple in a first-floor remodel on a slab foundation or a second-floor remodel in any type of house. If the drainage line you need to splice into is under the slab, then you need to get a concrete-sawing contractor to come and cut the floor so that the pipe can be excavated. It's not a good idea to break up the floor with a sledgehammer because the cracks in the concrete around the hole will be difficult to repair.

After the saw work and digging are done, laying in drains is usually easier (albeit more expensive) than doing it before the slab is poured because of the level walking surface around the cuts. For burial, some codes want water lines 1 ft. above and 1 ft. over to one side from soil lines. If you need to use the same trench to bring the water over to your new fixture location and do not want to make or cannot make that big of a trench, you'll need to get the local authority's approval for having the water closer to the drainage line. In cemented ABS or PVC, which rarely leak at joints, the inspector will probably let you off the hook.

The new drain lines should be in earth, under a vapor barrier to protect them from the new concrete that will be poured. You may not be able to get as good a fall on the drain line as on the original drain, and some of the new drain and/or water lines may have to be in the slab repour. This aspect of the work is also likely to need the local authority's approval, and you may be required to sleeve the new piping in plastic to allow for thermal expansion.

If you are running new drainage for a drastically different second-floor remodel on a slab house, you might find that bringing the upstairs drain down the outside (and boxing it in) and joining the building sewer outside in the ground is a better solution than a main drop or drops descending through an interior chase, which might require some floor demolition for a building-drain splice or some foundation demolition to get outside for a hookup to the existing building sewer. A drain on the exterior of the building is not a good solution for cold climates, however.

For second-floor remodels on an off-the-ground structure, sometimes the new bathroom drop(s) can be made in closet corners; then after poking through the first floor, you can swing-fit a combo or long sweep 90° elbow and head off to the building drain. Second-floor joist directions more or less dictate what floor-plan changes can occur as far as plumbing is concerned. Many second-floor remodel jobs can't be plumbed without adding some framing, mostly because of opposing floor joists that must be bored for new piping. You will have to use a lot of 3-in. closet and main drains on second-story work due to shallow or bored joists.

CUTTING IRON PIPE

Old "corked" iron pipe (poured lead in bell and spigot fittings) is difficult to work with. It is often far from being round. The belled iron may have a pronounced ridge seam from the casting process along part or all of the length of the pipe. There may also be raised letters on the pipe. Any of these conditions can interfere with the fit of a no-hub or Mission coupling.

Cutting old iron pipe isn't easy. Old, horizontal iron pipes are hard to cut cleanly; sometimes the cutter's chain merely crushes the bottom of the pipe. Old, vertical iron stacks running between floors can be very brittle and can splinter when you try to snap-cut them. They get this way because water doesn't sit in a vertical pipe, and so they dry and harden. If the spot where a wye, upright wye or combination wye needs to be spliced in leaves barely enough room between other fittings, or maybe between a bell and a fitting, you might not want to risk the snap cutter at all and instead use a reciprocating saw or a grinder with a carborundum blade to cut the section out of the line. You should use a variable-speed reciprocating saw, like the trigger-control models made by Milwaukee and Black & Decker. With two-speed or preset-speed saws, even the slow speed is often too fast for this procedure and your blades will be ruined in short order. You might need three blades or more (at $5 to $10 each) to cut the pipe. But if you crack a pipe, you'll have even more work to do, possibly including cutting back branch lines and adding fittings if you crack the main drain with the snap cutter.

SPLICING NEW PIPES TO OLD

Splicing into existing drain lines is a lot simpler than it used to be, thanks to rubber couplings. These couplings, which come in various configurations and are made by several companies, can be used for joining pipes of dissimilar sizes and materials. For more on these couplings, see pp. 90-93.

Splicing into existing no-hub DWV is relatively easy because you can simply undo no-hub couplings and remove sections of pipe. No-hub pipe tends to be truer in roundness than bell and spigot pipe so the couplings seal much better. But, sometimes the bands on no-hub couplings are installed so that the screws on the clamps are facing away from you and you cannot get at them. If this is the case, you must cut into the pipe, just as for old bell and spigot pipe.

If you plan on using a Mission coupling, you will have to cut out a piece about ¼ in. to ⅜ in. longer than the fitting you are splicing in. If you don't, the stop in the rubber no-hub sleeve may not fit between

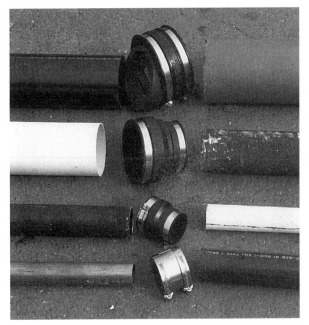

Dissimilar piping materials may be joined with a code-approved rubber coupling. From top to bottom: 4-in. ABS to 4-in. terra-cotta (a 4-in. Calder coupling with a bushing for ABS), 4-in. PVC to 3-in. iron (a 4-in. plastic to 3-in. iron Fernco coupling), 2-in. iron to 1½-in. PVC (a 2-in. iron to 1½ in. Fernco coupling), 1½-in. copper to 2-in. ABS (a 1½-in. copper to 2-in. plastic Mission coupling).

You can take the wrapping out of a Mission coupling clamp (left) and use it inside a no-hub clamp (right) for a tighter fit.

the two pieces and there will be a lump in the center of the rubber sleeve that will prevent the outer stainless cover from making a good seal when you tighten the screws. If you have to cut where there are raised letters on the pipe, you can use the grinder to remove them before cutting.

If you are splicing into bell and spigot iron pipe, you might find that the outside diameter of the old belled iron pipe is a bit smaller than the new no-hub iron fitting you are splicing in. Sometimes this difference in diameter is enough to keep a standard no-hub coupling from sealing properly, and the splice may leak. For this reason, you might want to use a Mission coupling instead of a standard no-hub fitting. The Mission coupling has a rubber sleeve with a smaller-diameter opening on one side of the stop for the old, corked iron. When using it on no-hub fittings and belled iron pipe, put the larger side (thinner rubber) on the no-hub fitting and the smaller side (thicker rubber) on the old belled iron pipe. When I use this technique, I take the rubber sleeve out of the standard stainless-steel sleeve (which is a draw-bolt type) and replace it with a worm-drive clamp from a standard no-hub clamp (see the photo above). This allows me to tighten it more, thus reducing the chance of a leak. (I should add that this is one of those very rare times when I do not go out of my way to share a trade secret with the local authorities.) I also use this

technique when beginning a run of ABS or PVC to a freshly cut belled iron pipe. In this case, however, I leave the Mission coupling taped to the pipe. If the inspector insists on the Mission draw-bolt clamp, I'll put it back on, but only after the water test.

When I cut open an iron DWV line for a splice, I install only an iron fitting. I do this because the possible shifting forces transmitted by iron pipe would put too much stress on a plastic fitting. I am not aware of any code requirements that it be done this way, but most local authorities will require it anyway. Off the branch of the spliced-in iron fitting, I will adapt to plastic or copper pipe.

Before you start cutting into iron pipe, try to ascertain if the piping is going to drop or sag when the piece is removed. (Supporting iron pipe is discussed in detail on pp. 116-118.) Sometimes on a vertical pipe, the opening you make can close up by as much as several inches, enough to prevent a wye or combo from fitting in when the cut piece is removed. On a horizontal run, one side of the cut pipe can drop so far that the fitting will not be flat when installed. It's a fairly easy process to hang the horizontal pipe on both sides of the cut (before you make it) with plumber's tape.

The vertical run is more of a challenge to support. If you are lucky and have some branched fittings on it below the floor joists but above your splice, you should wrap plumber's tape around and under the branch and then screw the plumber's tape as high up on the joist as you can with a cordless screwdriver (see the top drawing on the facing page). To support a vertical no-hub pipe you can also use a no-hub riser clamp (see the photo on p. 34), with strap steel or heavy-weight galvanized plumber's tape attached at the bolts or with blocking nailed to the joists on either side of the clamp. Just to get the fitting in place so drains may again be used, I can sometimes use 2x4s as stiff legs from the ground up to a branched fitting. Once the spliced fitting is in place on a vertical drop and the bands are tightened up, I remove the 2x4s and add some tape support. The compression on the fitting in a vertical drop will not usually cause any or as much misalignment (if all edges are close to flush) as the different elevations of a severed horizontal drain. Also, the larger the horizontal drain, the less truing effect the couplings will have when you tighten them up.

SUPPORTING VERTICAL IRON PIPE

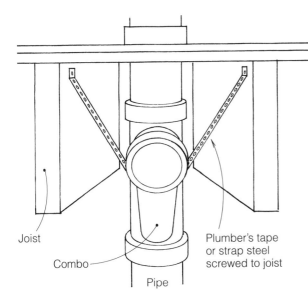

Joist

Combo

Pipe

Plumber's tape
or strap steel
screwed to joist

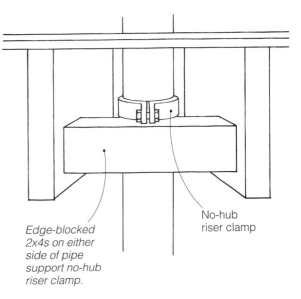

*Edge-blocked
2x4s on either
side of pipe
support no-hub
riser clamp.*

No-hub
riser clamp

SPLICING INTO PLASTIC PIPE

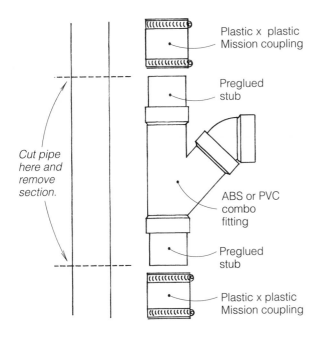

Plastic x plastic
Mission coupling

Preglued
stub

*Cut pipe
here and
remove
section.*

ABS or PVC
combo
fitting

Preglued
stub

Plastic x plastic
Mission coupling

Splicing into an ABS or PVC plastic line (see the drawing at left) is much easier than splicing into iron. All you need are plastic by plastic Mission couplings and a handsaw or reciprocating saw to cut the pipe. If you can force the severed drain line apart far enough to slip the branched fitting in, then you could cement in a plastic fitting (see pp. 54-58), but unless you have help to shove the pipe back into the fitting, I don't recommend doing so. If water continues to dribble out of the pipe, it might interfere with the cement's ability to make a watertight joint. And with plastic cement, you have just a few seconds to position the branch of the fitting; if you don't get it right, you'll have to cut out the fitting and begin again.

SEWAGE EJECTORS

If Mother Nature can provide the energy to drain wastes from the remodel, so much the better. But if the remodel is at basement level and wastes have to flow up to the level of the building sewer, then you'll need a sewage ejector. A sewage ejector is an electrically powered pump designed to pump solid and liquid waste up out of a sump (holding basin) and into the building sewer. Installing a sewage ejector requires the services of an electrician, and the initial cost of the equipment isn't cheap — about $900 for an average-sized sewage ejector.

The pump is located inside a sump that connects to a drain and a vent. The pumps and sumps for ejectors are sized for the different flow rates that they need to pump. Make a list of your fixtures and the fixture-unit total, and shop for an ejector that will comfortably fill your needs. Don't get one that is only marginally acceptable.

Sewage ejectors should be used only when there is no other solution. The sump is a bulky container that requires a large-diameter hole about 3 ft. deep that is difficult to integrate into a plan indoors. Because of the need to service and repair the system, it should have good access. It also needs to be served by electricity. Many codes want a switchable light right above the installation. This means hiring an electrician if you can't do the work yourself. I strongly recommend that you try every other plan possible before going this route and forever being subject to power outages and pump failures.

There are two basic types of sewage ejector: automatic and manual. The automatic system has a pump with a mercury float switch (a barrel-shaped float with mercury inside) on a short cord wired directly into the pump motor. When the float is hanging down, as it would in an empty sump, the mercury is at the bottom of the float, away from the two wire contacts that when immersed in mercury will complete the circuit. When the sump fills up, the float rises until the mercury surrounds the contacts and the pump is activated. For this design, the pump itself (in most cases) is just a bigger, much more powerful version than would be used for water only.

In the manual design, the pump and the switch are separate. The switch rests on top of the sump's lid and remains dry. An electrical cord from the motor in the bottom of the sump plugs directly into the switch. The switch has a small braided cable that hangs down into the sump. On this cable are secured two or more weighted floats that sense the rising liquid level. When the sump is empty, the weighted floats pull down on the switch, keeping it in the off position. When the level in the sump rises, the floats become somewhat buoyant, and the switch turns on the pump. (Even though this system also operates au-

tomatically, where you have a separated switch, the pump is referred to as manual.) With this system, the plumber installs the switch separately and installs the weighted floats on the cable at suggested heights. It is sometimes necessary to tinker with the float adjustments to get the switch to activate at the proper time.

In my experience, automatic sewage ejectors are less troublesome than manual sewage ejectors. They are also much quicker to install, but when the controls need replacing, the pump has to be lifted out of the sump, which means that the discharge piping and the lid need to be removed. On manual sewage ejectors, the controls can be replaced without removing the lid and the pump. If someone were to ask me which type would I recommend, however, I would say the automatic with mercury float switch.

Whichever type you end up using, you'll have to connect the sump to the building drain. The plastic sumps I've seen lately leave a lot to be desired in terms of the connection to the building drain. One type has a rubber sleeve designed like a grommet that is supposed to fit in the wall of the sump in a little circular groove and then grip the pipe with a plastic cinch strap. I don't like this design because I think the cinch strap is too weak. Another type has a boss or sleeve on the outside for the pipe to slip through. Between the outside edge of the pipe and the inside wall of the sleeve, you install a rubber bushing. When I use this type, I don't rely only on the bushing. I find some form of no-hub rubber coupling like Fernco, Calder or Indiana Seal that I can adapt if necessary to grip the outside of the sleeve and the pipe as well. (For more on these couplings, see pp. 90-93.)

From the sump, you have to run a vent (2-in. dia. minimum) that must terminate above the roof just like any other vent. You also have to install a union on the discharge pipe, above the lid but below the check valve and gate valve, to facilitate removal of the pump. A check valve allows the flow in a pipe to go in only one direction (in this case away from the pump). The check valve keeps the pump from working continually and burning out. Most codes want a check valve and a shut-off valve (I use a full-port ball valve) close above the sump. For all of these components, use the best-quality materials you can find — a sewage spill is not a pleasant thing.

SEWAGE EJECTORS

AUTOMATIC PUMP

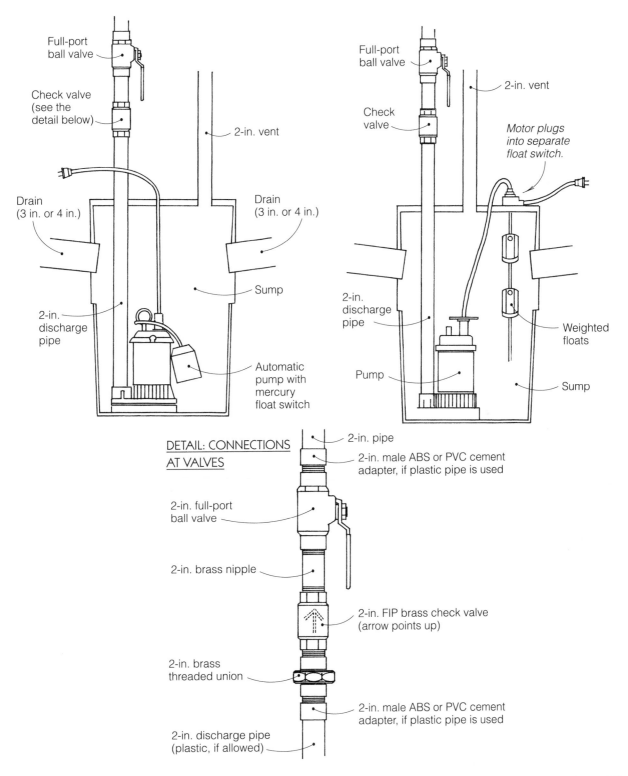

Full-port ball valve

Check valve (see the detail below)

2-in. vent

Drain (3 in. or 4 in.)

Drain (3 in. or 4 in.)

Sump

2-in. discharge pipe

Automatic pump with mercury float switch

MANUAL PUMP

Full-port ball valve

2-in. vent

Check valve

Motor plugs into separate float switch.

2-in. discharge pipe

Weighted floats

Pump

Sump

DETAIL: CONNECTIONS AT VALVES

2-in. pipe

2-in. male ABS or PVC cement adapter, if plastic pipe is used

2-in. full-port ball valve

2-in. brass nipple

2-in. FIP brass check valve (arrow points up)

2-in. brass threaded union

2-in. male ABS or PVC cement adapter, if plastic pipe is used

2-in. discharge pipe (plastic, if allowed)

If you are using ABS or PVC piping, there are cement adapters with male pipe threads that will allow you to thread to the check valve and shut-off valve (see the detail drawing on p. 199). I install a union below the shut-off valve so I can get into the sump without having to cut any pipe. I use cemented male iron pipe (MIP) adapters here again so I can use a brass union, which will always be easy to undo (not so with galvanized steel). I don't use plastic unions or no-hub couplings, because I want the best protection I can have against leaks. No-hub couplings are not designed for pressure systems.

For the shut-off valve, get a full-port ball valve, not a gate valve, because it's easy to tell from a distance whether the lever-handled ball valve is open or closed. It is almost impossible to tell from a distance whether a gate valve is open or closed, and the consequences of the pump working against a closed valve can be disastrous. If you need to service the pump and then forget to open the valve before the pump cycles, the "dead head" condition can destroy the pump by blowing its seals. Many companies manufacture valves. The two foreign valves I like are Red and White, from the Toyo Valve Co. (Japan) and the Pegler line, made in England. Otherwise I stick to reputable American lines like Crane, Grinnel, Hammond, Honeywell, Milwaukee, Nibco and Stockham. Few home owners understand what anguish and expense they are risking with a cheap valve until it malfunctions and leaks.

Where you place your sewage ejector can be an important decision, and your local authority might have something to say about it. Check with your inspector before you invest a lot of energy on an unacceptable location. It's usually best to put a sewage ejector outside the house. However, an outside location may not be possible because of property lines or other space limitations. If you need to locate the ejector inside the building, many locales will require that you place it inside a concrete vault as secondary containment in the event of spills.

FRESH-WATER SUPPLY

If you need to bring water into an addition, you will need to choose a piping material, size the system and assemble the pipes. Your choice of materials will probably be influenced by the local standards. If you are in an area where PB or PE is allowed, review pp. 63-64 to see if they are materials which you might like to work with. PVC is another plastic option. For a general discussion of piping materials suitable for water supply, see pp. 29-33.

Also, you should think about the proper pipe sizing for your new work. Even if the existing system is undersized to current code, you should do your new work to code. If you can choose splice locations that afford access even after the new work is completed, the local authority will sometimes not demand that you bring up the rest of the house to code before moving on with your project. For more on sizing water-supply piping, see pp. 45-49.

New piping needs protection from nails and screws. Devices called nail plates (see p. 141) are nailed over studs and joists to protect the piping from the drywall installation and finish carpentry. Plumber's nail plates are not as strong as those sold at electrical-supply stores, but they are better than nothing at all. Install nail plates on all wood members when the pipe is within penetration distance of nails and screws.

If you are about to embark on a remodel of an existing space, you can end up boring almost equal distances through new and existing studs. It's the existing wood that's a problem due to nails, especially the studs on exterior walls, where you are liable to hit 8d nails and sinkers from the studs and the blocking. Plumber's boring bits are expensive, and repeated encounters with nails can destroy them. Try to select plumbing pathways with the fewest embedded nails. Your progress will be much more rapid because you won't have to use a hole saw (see p. 11), which cuts more slowly than a drill bit.

Lenox makes excellent small-diameter boring bits, but their shanks are too short for boring double or triple studs or a channel, so you need to use an extension. Plumbers use right-angle drills like the Milwaukee Hole Hawg because the tight right angle affords more room to fit the longest drills possible between studs. To get through double or triple studs, you might have to start with the smallest extension and work up to a longer one, or drill from both sides.

NOTCHING VS. BORING

Sometimes notching the studs is a better alternative than boring them. I think that bored holes make for a stronger wall, but in impossible-to-bore situations (for example, where diagonal bracing fills the space on one or both sides of the studs), notching is the way to go. Never notch deeper than the diameter of the pipe, but if you have been boring holes down the stud wall, you are probably in deeper than what the usual notch depth is. In this case, you might use a female by female 45° elbow with a street 45° elbow back far enough from the notch to bring the pipe out to notch depth and then use two more of the fittings to return to the holes on the other side, as shown in the drawing below. Try to avoid seriously weakening structural members.

SPLICING COPPER PIPE

If you need to splice into existing copper tube to supply water on a remodel, the use of the copper repair coupling (see p. 74) will serve you well. As long a you do a very thorough cleaning job on the pipe and coupling, you should have few problems. If water in the pipe is a factor, sopping it up with French bread (see p. 76) should help alleviate the problem. Remember that when soldering there is the risk of starting a fire, so make sure you have on hand a squirt bottle, a small piece of sheet metal or fabric flame-guard material, and a fire extinguisher.

If you have existing metal water-supply piping (galvanized steel or copper) and you are planning to attach plastic pipe to it, you should check with an electrician to make sure that you do not isolate any grounds and endanger the occupants electrically. Also check with an electrician if you have existing galvanized-steel pipe and are going to splice into it and install copper tubing for the addition, since dielectric unions can also isolate grounds. If you need to have an electrician install a new ground to any isolated copper pipe, make sure that the grounding clamp is solid brass, including the screws. If a plated-steel one is used, electrolysis can destroy the ground and the pipe.

NOTCHING STUDS

TOP VIEW

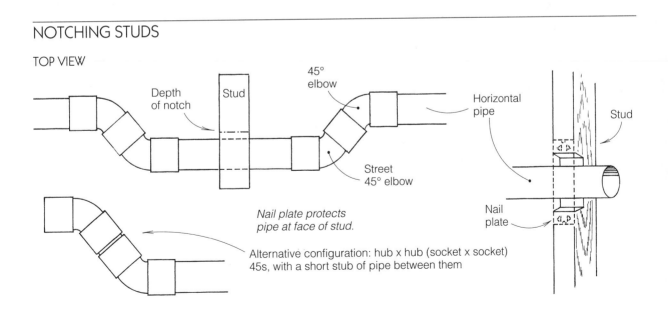

Depth of notch
Stud
45° elbow
Street 45° elbow
Horizontal pipe
Stud
Nail plate

Nail plate protects pipe at face of stud.

Alternative configuration: hub x hub (socket x socket) 45s, with a short stub of pipe between them

A dieletric union (left) and dielectric coupling (right) are useful in remodel work for joining steel pipe to copper.

For attaching different materials to copper pipe, there are several options. For attaching galvanized-steel pipe, you can use dielectric unions or dielectric couplings (see the photo above), or 6-in. long brass nipples. Dielectric unions are threaded-steel unions, half of which threads onto the pipe and half of which is a copper bushing, into which the copper tube is soldered; a rubber washer and plastic sleeve inside the coupling isolate the copper from the steel. Dielectric couplings are steel with an inner threaded plastic sleeve. A steel nipple or a steel male-threaded fitting is threaded into one side; a threaded copper to sweat adapter, with copper pipe soldered to the adapter, is threaded into the opposite end. The threaded plastic sleeve isolates the outer steel sleeve from the copper, and the coupling is long enough that the steel or iron fitting and the copper adapter never touch each other. Long brass nipples are threaded into steel female fittings, which are threaded onto steel pipe. Onto the opposite end of the brass nipple, a threaded female copper to sweat adapter is threaded. Copper pipe is soldered into the back of the copper adapter. The brass nipple resists the dielectric action of the copper and steel at each end.

For attaching PB to copper, a copper barb to sweat adapter is soldered onto the copper pipe, and then the barb end is crimped in the PB pipe. For attaching PE to copper pipe, you can use a brass barb to brass FIP adapter. Into the FIP adapter you thread a copper male to copper sweat adapter. Into the sweat half of the copper adapter, you solder the copper pipe. (The soldering is done before connecting the plastic pip-

ing.) For attaching PVC to copper, a plastic MIP to slip is threaded into a copper FIP to sweat adapter (see the photo on p. 62). A sweat, wrought-brass, copper to copper to FIP tee would provide a threaded metallic female opening on a straight run of copper pipe for which to thread in a threaded PVC adapter for continuing an intersecting PVC run of pipe.

SPLICING GALVANIZED-STEEL PIPE

If you need to splice into an existing galvanized-steel water line, you will have to cut some threads. To my knowledge, there is no other acceptable option that is allowed to be buried in walls. Old galvanized-steel water lines will usually thread without much problem if they are off and out of the ground and in walls, attics or hung from joists. In fact, most 50-year-old American-made galvanized pipe threads better and more easily than most new imported pipe. You just need a work area that affords enough space for the tools and for attaching the fittings.

Inside a wall, for water supply only, most locales allow the use of threaded unions. These are much easier to install than right/left couplings and nipples (see pp. 173-174). But I wouldn't be surprised if some locales demanded right/left couplings and nipples for water lines inside a wall (most states require them for gas lines). If you are allowed a union and you are working vertically, through a hole in an existing wall, use the half of the union that holds the union nut on the top of the joint. If you use it on the bottom, you may drop the nut, and it could easily slide beyond reach. One union, one branch fitting and a nipple of any length are the fittings needed to splice into a threaded-steel line.

As mentioned earlier in the book, olive oil is a better thread-cutting lubricant than petroleum-based thread-cutting oils. In remodel work especially, you and your client will appreciate olive oil's lack of odor and ease of cleanup (soap and water). For a detailed description of threading galvanized-steel pipes, refer to pp. 82-83.

If you are under a first-floor crawl space or a full basement making splices into existing water lines or gas lines, you should install a valve close to your newly installed tee's branch. This valve allows you to resume service to the rest of the house quickly and also to isolate the new plumbing system from the existing plumbing system for future service and repair. Valves in this application are called isolation valves. Most valve companies make a WOG (water, oil and gas) valve (see the photo at top left on p. 167). These valves come in handy because you can use them for any of the three common substances.

Sometimes you have to splice in tight quarters, and you might not have enough room to swing the handle on your threading ratchet. The handle will bang into one side of a wall opening or structural member before the ratchet clicks into the next slot in the die, rendering the tool useless. In situations like this, I have had to use my 18-in. pipe wrench on the die by itself without the ratchet to cut the needed threads. Remember to back up the pipe that is being threaded with a pipe wrench (apply the wrench to the pipe being threaded to oppose the torque applied by the threading head) or you can break a joint farther or deeper in the wall or floor. If that happens, a little job turns into a huge job.

FUEL-GAS LINES

If you need to bring a gas line to a new location in the remodel or addition, you might need to increase the size of the pipe. Since boring walls full of other pipes, ducts and wires and assembling a new line out of numerous short pipes, couplings, nipples and elbows is quite a hassle, you might find it a lot easier to run a completely new line from the meter.

If you decide to battle it out in existing walls, false ceilings and attics, you'd be wise to complete the run in successive short sections and test them as you go. If you use a WOG ball valve for your gas isolation valve near the spliced-in tee, it will withstand whatever pressure the local authority wants you to test at. If you do a long line with lots of fittings and then have a leak in the middle, you will have to take everything apart from one direction until you get to the leak, repair it and reassemble the pipes, possibly creating another leak in the process. Otherwise you have to saw through the pipe, fix the leak and then use a right/left nipple and coupling (see pp. 173-174) to rejoin it.

Threaded unions for gas are not allowed under homes or inside walls in most codes. Regardless of what some helpful tradesman, friend or neighbor might advise, use both Teflon tape on the male threads and pipe dope in the fitting and you'll greatly reduce the chance of leaks. Never use the merchant's coupling that is shipped threaded to the end of the pipe. Its walls are thin, and when you snug this type up onto a nipple or pipe, it can spread easily, causing a leak. Use a heavy-walled, forged coupling instead (see the photo on p. 166).

A WATER-HEATER UPGRADE?

If you are rearranging walls and fixtures within a structure whose square footage remains the same, you probably will not need an additional water heater, although here is your opportunity to upgrade the pipe size for both supply and delivery. If you are about to build an addition with hot-water requirements, whether it represents an additional floor or more space on the same floor, you should assess the additional demands that will be placed upon the existing system (discuss the situation with your local authority). More often than not, the existing piping is undersized. Even if your inspector would let you construct the addition without increasing the size of the main and the major supply line from the water heater, if the addition will host a tub or shower and it is physically possible, it is a good idea to run a ¾-in. pipe from the hot-water nipple on the heater all the way to the fixture.

The water-heater connection (see the drawing on the next page) is best made by stacking tees on the hot-water port of the heater instead of teeing into one existing line. Also put a full-port ball valve on the new hot-water supply line to the addition close to the water heater so that you can throttle the supply to the addition and adjust for pressure drop due to water use in other parts of the house (see p. 162); the valve will also let you turn off the hot water to the addition for maintenance and leave it on for the rest of the structure. If you can find a convenient spot for a valve on the cold-water supply line to the addition, it is a good idea to install one there, too.

CONNECTING NEW LINES TO THE WATER HEATER

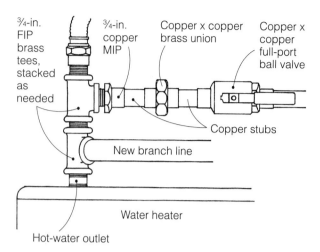

¾-in. FIP brass tees, stacked as needed

¾-in. copper MIP

Copper x copper brass union

Copper x copper full-port ball valve

Copper stubs

New branch line

Water heater

Hot-water outlet

Even if you had not considered the need for a hot-water circulating pump (see pp. 154-155), you might want to consider one now, when the walls are opened up. On far-flung tubs, showers and sinks, it's irritating and wasteful of resources to have to wait a long time for hot water. A circulating pump takes stored, heated water from your storage-tank water heater and keeps a small flow of water going in a ½-in. piped loop from the water heater to the distant fixture. When you turn on the hot water, you have hot water right away. You do not need to have this pump running 24 hours a day; it can be put on a timer or pressure switch to match your usual needs. If you decide after the project is completed that you want a circulating pump, the cost of going back and opening up all of the necessary walls will often be prohibitive. You can add the loop now by hooking into the hot-water branch at the fixture and capping off the end of the line at the water heater. Then later, if you decide that you do want a circulating pump, the most expensive part of the job is already done.

If you are doing a project in stages and know that down the road you will be moving the location of the water heater, plan for it now. Perhaps you have the space for a convenient water-heater closet or maybe a full-height storage space under the addition you are building. You can install a second water heater in the addition now and plumb it so that it will supply the house efficiently when the old heater is retired. You can install tees and valves at both heaters so that either one can be isolated from the system and can supply all the hot water to the structure. Try living with the old water heater first. If it works, you can mothball the new one (leave if off and empty) and wait until the next phase to remove the old heater and bring the new one on line. If your original plan of serving the addition with the old heater proves unsatisfactory and you opted not to put in a recirculating line, you can always run both heaters.

AN ELECTRIC WATER HEATER FOR THE KITCHEN OR BATHROOM

If you are doing a kitchen or bathroom remodel and chronically poor hot-water arrival time is a problem, you can have an electrician run a separate circuit to under the kitchen cabinet, where you could install a small (1-gal. to 3-gal.) holding-tank electric water heater or a small tankless water heater to provide immediate hot water until the hot water from the main supply arrives. The circuit needs to be properly rated for the demands of the chosen heater. Most of these heaters are 110 volt. I like a brand called Ariston and another version by the Insinkerator Company.

As shown in the drawing on the facing page), you hook up the heaters by pulling the existing hot water (from the hot angle stop) into the cold inlet of the heater and take the hot-water line from the heater out and up to the hot connection for the faucet.

With the Ariston, there are two possible configurations, depending on the distance from the heater to the hot-water faucet's connection. If the heater and the faucet connection are less than 30 in. apart, you can use a generic stainless-steel braided supply or Aquaflo's ½-in. FIP by ½-in. FIP by 30-in. long, rubberless faucet supply to make your connections. You install a new ½-in. FIP by ½-in. IPS angle stop on the sink's hot-water supply nipple, then use a brass ½-in. street 90 on the angle stop and on the water-heater nipples.

If the distance between connections is too great to be served by the Aquaflo components or generic stainless-steel braided supply, you can use copper refrigeration tubing instead. Install ½-in. FIP by ⅜-in. compression angle stops and then a ½-in. FIP by ⅜-in. compression adapter on the hot-water faucet con-

CONNECTING AN ELECTRIC WATER HEATER TO THE KITCHEN-SINK HOT-WATER FAUCET

ARISTON WITH AQUAFLO OR STAINLESS-STEEL BRAIDED SUPPLY

INSINKERATOR WITH AQUAFLO OR STAINLESS-STEEL BRAIDED SUPPLY

ARISTON WITH COPPER REFRIGERATION TUBING

INSINKERATOR WITH COPPER REFRIGERATOR TUBE

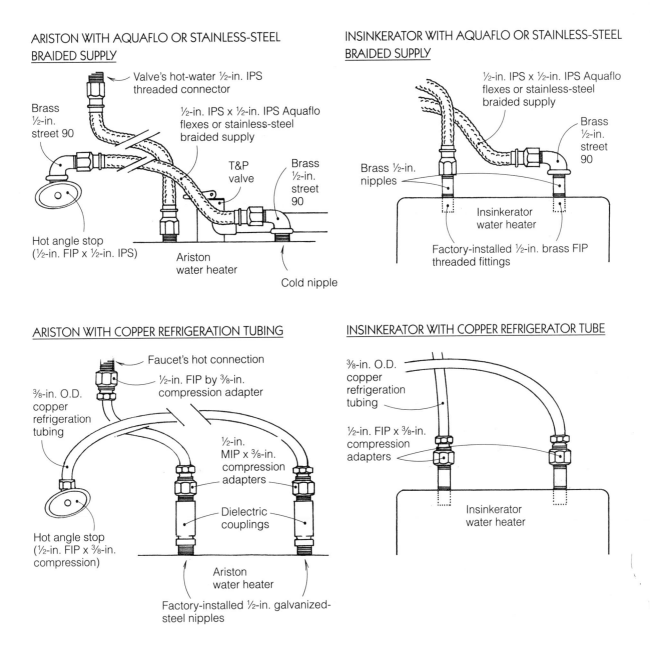

nection. Put dielectric couplings and ½-in. MIP by ⅜-in. compression adapters on the water heater's ½-in. steel male pipe nipples and then use ⅜-in. O.D. copper refrigeration tubing for the supplies. This is the same size as standard ⅜-in. faucet supply stock anyway. The hot-water angle stop acts as a shut-off valve for the water heater, and the brass ferrules keep the setup rubberless.

With the Insinkerator, the connection at the water heater is slightly different. Use ½-in. by 2-in. brass nipples in the heater, onto which you install a ½-in. FIP by ½-in. IPS angle stop with ½-in. FIP by FIP Aquaflo or braided supply supplies up to the faucet.

SOURCES OF SUPPLY

ACE CORP.
2200 Kensington Court
Oakbrook, IL 60521
(708) 990-6600
FAX (708) 571-2186
Pipe-nipple backout; nipple extractor

**AMERICAN SAW AND
MANUFACTURING CO.**
301 Chestnut St.
East Longmeadow, MA 01028
(800) 628-3030
FAX (800) 223-7906
*Lenox self-feed drill bits; blades for
reciprocating saws*

AMES TOOL
P.O. Box 1774
3801 Camden Ave.
Parkersburg, WV 26101
(800) 624-2654
FAX (304) 424-3330
Water-heater vent-pipe fittings

AMTROL, INC.
P.O. Box 1008
West Warwick, RI 02893-1008
(800) 726-6962
FAX (401) 885-2567
*Residential pressure boosters;
expansion-tank and check-valve
systems*

**AQUAFLO
DIVISION OF SPECIALTY PRODUCTS**
3689 Arrowhead Drive
Carson City, NV 89706
(800) 854-3215
FAX (702) 884-4343
Braided stainless-steel supply tubes

ARISTON
Controlled Energy Corp.
(distributor)
Fiddler's Green
Waitsfield, VT 05673
(800) 642-3111
(802) 496-4436
FAX (802) 496-6924
*Electric water heaters; tankless water
heaters*

**CASPER'S INDUSTRIES
DIVISION OF OATEY**
6600 Smith Ave.
Newark, CA 94560
(510) 797-4672
FAX (510) 790-3442
Closet flanges

COLEMAN CO.
P.O. Box 2931
Wichita, KS 67201
(316) 261-3211
FAX (316) 261-3563
Illuminators

CRANE CO.
100 1st Stamford Place
Stamford, CT 06902
(203) 363-7300
FAX (203) 363-7295
Full-port ball valves

DE BEST MANUFACTURING
P.O. Box 646
Gardena, CA 90248
(310) 352-3030
FAX (310) 327-1921
*Cardboard sleeves for closet risers in
slabs; tub boxes*

**DIAMOND TOOL CO.
DIVISION OF COOPER GROUP**
Cameron Rd.
Orangeburg, SC 29115
(803) 533-7862
FAX (803) 536-6632
Diagonal-cutting pliers

ESTWING
2647 8th St.
Rockford, IL 61109
(815) 397-9521
FAX (815) 397-8665
Cat's paws

FERNCO
300 South Dayton St.
Davison MI 48423
(313) 653-9626
Rubber no-hub couplings

**FRANK PATTERN AND
MANUFACTURING, INC.**
508 Winmoore Way
Modesto, CA 95358
(209) 538-3251
FAX (209) 537-7562
Closet flanges

GRINELL CORP.
2175 15th St.
Denver, CO 80202
(303) 825-7181
FAX (303) 825-2321
Full-port ball valves

GROHE CO.
241 Covington Drive
Bloomingdale, IL 60108
(708) 582-7711
FAX (708) 582-7722
Temperature-compensating valves

HAMMOND VALVE CO.
210 Tower St.
Prairie du Sac, WI 53578
(608) 643-2977
FAX (608) 643-3775
Full-port ball valves

HERCULES CHEMICAL CO.
29 West 38 St.
New York, NY 10018
(212) 869-4330
Pipe cements and primers; pipe-joint compound

HONEYWELL
2701 4th Ave. South
P.O. Box 524
Honeywell Plaza
Minneapolis, MN 55408
(612) 951-1000
FAX (612) 951-2086
Full-port ball valves

HUBBARD ENTERPRISES
303 Enterprise St.
San Marcos, CA 92069
(619) 744-6944
FAX (619) 744-0507
Holdrite pipe-support brackets

IDEAL CORP.
3200 Parker Drive
St. Augustine, FL 32095
(904) 829-1000
FAX (904) 829-6994
Rubber no-hub couplings

INDIANA SEAL CO.
12000 Mitchel Rd.
Carmel, IN 46032
(800) 428-5309
FAX (800) 422-8195
Rubber no-hub couplings

INSINKERATOR CO.
4700 21st St.
Racine, WI 53406
(414) 554-5432
FAX (414) 554-3639
Electric water heaters; tankless water heaters

INSULATION MATERIALS CORP. OF AMERICA
4325 Murray Ave.
Haltom City, TX 76117
(817) 485-5290
FAX (817) 656-5394
Imcoa K insulation

JOINTS, INC.
311 South Highland Ave.
Fullerton, CA 92632
(714) 525-1613
FAX (714) 525-6469
Calder couplings

KLEIN TOOLS
7200 McCormick Blvd.
P.O. Box 599033
Chicago, IL 60659-9033
(708) 677-9500
FAX (708) 677-4476
Aviation snips

MALCO PRODUCTS
14080 State Highway 55 NW
P.O. Box 400
Annandale, MN 55302-0400
(612) 274-8246
FAX (612) 274-2269
Bit-tip sheet-metal screws; hand punch

MILL ROSE CO.
7995 Tyler Blvd.
Mentor, OH 44060
(216) 255-9171
FAX (216) 255-5039
Pink Teflon tape

MILWAUKEE ELECTRIC TOOL CO.
13135 West Lisbon Rd.
Brookfield, WI 53005
(414) 781-3600
Hole Hawg right-angle drill and drill bits; Sawzall reciprocating saws

MILWAUKEE VALVE CO.
2375 Burrell St. (Dept. TR)
Milwaukee, WI 53207-1519
(414) 744-5240
FAX (414) 744-5840
Full-port ball valves

MISSION RUBBER CO.
P.O. Box 2349
Corona, CA 91718
(909) 736-1881
FAX (909) 736-0481
Mission brand couplings; band seals; all-rubber fittings

MOEN
25300 Al Moen Drive
North Olmsted, OH 44070
(216) 962-2000
FAX (216) 962-2770
Temperature-compensating valves

**MUELLER
DIVISION OF STREAMLINE**
555 N. Woodlawn
Wichita, KS 67208
(316) 688-8100
FAX (316) 688-8182
Copper pipe and fittings

NASHUA CORP.
P.O. Box 3000
Merrimack, NH 03054
(603) 880-1111
FAX (603) 880-1251
Solid aluminum tape

NIBCO
500 Simpson Ave.
Elkhart, IN 46515
(219) 295-3000
FAX (219) 295-3307
Copper pipe and fittings; full-port ball valves

NOKORODE
DIVISION OF THE DUNTON CO.
3 Bridal Ave.
West Warwick, RI 02893
(800) 556-6538
FAX (401) 821-1914
Lead-free paste flux

NORTON
P.O. Box 549
Corona, CA 91720
(909) 278-2365
No-hub couplings

OATEY CO.
4700 West 160 St.
P.O. Box 35906
Cleveland, OH 44135
(216) 267-7100
FAX (800) 321-9535
Pipe cements and primers

OTC
855 Eisenhower Drive
Owatonna, MN 55060
(507) 455-7000
FAX (507) 455-7300
Pry bars

PAASCHE
7440 West Lawrence Ave.
Dept. 99B
Harwood Heights, IL 60656
(800) 621-1907
(708) 867-9191
FAX (708) 867-9198
Compressors for testing fuel-gas systems

PASCO
11156 Wright Rd.
Lynwood, CA 90262
(310) 537-7782
FAX (310) 537-5065
Plumbing tools; mercury test gauges

PERFECTION PRODUCTS CO.
P.O. Box 40
Waynesboro, GA 30830-0040
(404) 554-2101
Stand-up wall furnaces; through-the-wall safety-vent space heaters

PIPE TYTES
100 Industrial Park Drive
Boone, NC 28607
(704) 264-2726
FAX (704) 262-0963
Insulators and fasteners for copper and plastic pipe

PORTER-CABLE CORP.
P.O. Box 2468
Jackson, TN 38302
(901) 668-8600
FAX (901) 664-0525
Cordless drills

PROTO TOOL
Stanley-Proto Industrial Tools
14117 Industrial Park Blvd.
Covington, GA 30209
(404) 787-3800
FAX (800) 438-9702
Socket wrenches

QUEST CO.
DIVISION OF U.S. BRASS
17120 Dallas Parkway
Dallas, TX 75248
(800) 872-7277
FAX (800) 525-7213
Q Talons plastic strap for hanging pipe

RECTORSEAL CORP.
2830 Produce Row
P.O. Box 14669
Houston, TX 77021
(713) 928-6423
FAX (713) 928-2039
Pipe-joint compound

REED MANUFACTURING CO.
1425 West 8th St.
P.O. Box 1321
Erie, PA 16512
(814) 452-3691
FAX (814) 455-1697
Copper tubing cutters

RESOURCES CONSERVATION
P.O. Box 71
Greenwich, CT 06836
(203) 964-0600
Scald Safe temperature-compensating faucet aerators, tub spouts and shower-arm regulators

RIDGE TOOL CO.
400 Clark St.
Elyria, OH 44036
(216) 323-5581
Ridgid brand plumbing tools

SANDVIK
P.O. Box 1220
Scranton, PA 18501
(717) 587-5191
Hole-saw indexes

SEARS, ROEBUCK AND CO.
3333 Beverley Rd.
Hoffman Estates, IL 60179
(708) 286-2500
Craftsman brand general tools

SELKIRK-METALBESTOS
17120 Dallas Parkway
Suite 205
Dallas, TX 75248
(214) 250-1795
FAX (214) 407-7261
Water-heater vent-pipe fittings

SIOUX CHIEF
P.O Box 397
Peculiar, MO 64078
(800) 821-3944
FAX (816) 758-5950
Water-hammer arrestors

SK HAND TOOLS
3537 West 47th St.
Chicago, IL 60632
(800) 822-5575
(312) 523-1300
FAX (312) 523-1300
Socket wrenches

**SPECIALTY PRODUCTS
DIVISION OF LPS PLUMBING
SPECIALTIES**
P.O. Box 1807
3689 Arrowhead Drive
Carson City, NV 89702-1807
(800) 854-3215
(714) 828-9730
FAX (800) 243-1777
*Sure-Sleeves brand through-form pipe
sleeves; plastic straps for hanging
pipe*

STREAMLINE
555 North Woodlawn
Wichita, KS 67208
(316) 688-8100
FAX (316) 688-8182
Copper pipe and fittings

TOYO VALVE CO.
20600 Canada Rd.
El Toro, CA 92630
(800) 222-7982
FAX (713) 859-7200
Full-port ball valves

TURBOTORCH
101 South Hanley St.
St. Louis, MO 63105
(314) 721-5573
FAX (314) 721-4822
Torches for soldering

U.S. BRASS
17120 Dallas Parkway
Dallas, TX 75248
(800) 872-7277
FAX (800) 525-7213
Copper pipe and fittings

WATTS CO.
815 Chestnut St.
North Andover, MA 01845-6098
(508) 688-1811
FAX (508) 794-1848
*Water-hammer arrestor; pressure-
reducing valves*

WHEELER MANUFACTURING
P.O. Box 688
Ashtabula, OH 44004
(800) 321-7950
FAX (216) 992-2925
Soil-pipe cutters

WILLIAM HARVEY CO.
4334 South 67th St.
Omaha, NE 68117
(800) 228-9631
FAX (402) 331-2118
Tub boxes

THE WILLIAMS CO.
225 Acacia St.
Colton, CA 92324
(800) 266-0993
(909) 825-0993
FAX (909) 824-8009
*Stand-up wall furnaces; through-the-
wall safety-vent space heaters*

WORLD WIDE WELDING
Nasco Inc. (distributor)
4749 Old Highway 8
St. Paul, MN 55142
(612) 780-2000
FAX (612) 783-2210
Lightnin' Bug torch lighter

INDEX

A

ABS (acrylonitrile butadiene styrene):
   advantages vs. disadvantages of, 22-23
   cements for, 55, 60-61
   cutting, 54
   development of, 22
   as DWV piping material, 21, 22-23
   fittings for, 23
   foam-core, 22
   joining, 54-57
   and transitional joint to dissimilar materials, 62
   weights of, 23
Airbrush compressor, plumbing uses for, 175
Altitude:
   and building-drain layout, 96-97
   defined, 96-97
   and selection of fittings, 98
Angle stops, anchoring for, 152

B

Bathtubs. *See* Tubs.
Bell reducers, for in-wall gas-pipe connections, 172
Bends. *See* Elbows.
Bicycle pumps, plumbing uses for, 129
Bidets, vent-to-drain patterns for, 105

Bits:
   auger, plumbing uses for, 11
   extension, plumbing uses for, 12
   self-feed, plumbing uses for, 11-12
Branch lines:
   defined, 28
   sizing, 48
Brushes, for cleaning copper fittings, 65
Building drain:
   and branch-drain connections, 99
   and cleanout locations, 100
   and connection to building sewer, 127-128
   and connection to drains in remodels, 194
   defined, 15
   design of, 98
Building sewer:
   and aggregate vent area, 52
   defined, 15
   sizing, 52
Buttonbead, as gas-venting material, 176

C

Cat's paws, plumbing uses for, 2
Cements:
   for ABS, 60-61
   cure times of, vs. ambient temperature, 61
   fumes from, 54
   for PVC, 62
Chalk lines, plumbing uses for, 4

Chisels, plumbing uses for, 3
Circular saws, plumbing uses for, 89
Cleanouts:
   defined, 21
   Kelly fittings as, 21, 127-128
   for sinks, 106
   on slab, 133-134
   tees as, 21
Closet flanges. *See* Fittings.
Combination square, plumbing uses for, 4
Compasses, plumbing uses for, 4
Copper:
   advantages vs. disadvantages of, 23, 30
   cutting, 64, 69
   and dielectric corrosion, 30
   as DWV piping material, 21, 23-24
   flare joints for, 72-73
   pipe support for, 34-35, 36-38
   repairing leaks in, 74-77
   rolled vs. rigid, 29-30
   soldered joints for, 69-71
   splicing, in remodels, 201-202
   uses for, 23
   and water hammer, 30
   as water-supply piping material, 29-30
   weights of, 23, 29
Couplings:
   for ABS and PVC, 19-20
   all-rubber, 92-93
   Calder, 94, 95
   CIT, 20, 93
   for copper, 19-20
   defined, 19
   for dissimilar materials, 195-196, 202

liquefied petroleum (LP) vs.
    natural, 164-165
piping materials and fittings
    for, 165-166
sizing, 169-171
vents for, 176-191
See also Gas-distribution piping.
Furnaces, stand-up wall, 190-191

G

Galvanized steel:
    advantages and disadvantages
        of, 30-31
    assembly of, 84-86
    cutting, 78-81
    pipe support for, 34, 36-38, 167
    splicing, in remodels, 202
    threading, 82-83
    as water-supply piping material,
        30-31
Gas appliances:
    fuel demands of, 170
    vent-piping materials for,
        176-177
Gas-distribution piping:
    discussed, 164-165
    hangers for, 167
    layout of, 171
    in remodeling, 203
    repair of, with right/left
        couplings and nipples,
        173-174
    sizing of, 169-171
    See also Fuel-gas systems.
Gas-venting systems:
    described, 176
    for dryers, 191
    pipe support for, 180
    piping materials for, 175-176
    roof flashing for, 180
    for stand-up wall furnaces,
        190-191
Grinders, plumbing uses for, 89

H

Hacksaws, plumbing uses for, 8
Hammer arrestors, commercial,
    144
Hammers:
    demolition, plumbing uses for,
        10
    plumbing uses for, 2
Hand punches, plumbing uses for,
    186
Hangers. *See* Pipe support.
Head pressure, and water-supply
    system design, 44
Headlamps, plumbing uses for, 13
Hose bibbs:
    with Jim cap, 130
    standard vs. freeze-proof, 157
    as washing-machine supply
        valves, 157
Hot-water dispensers, water-supply
    piping for, 151

I

Illuminators, plumbing uses for,
    13
Insulation:
    extruded closed-cell foam, 39
    fiberglass wrap, 39
    for water-supply piping, 139
        under slab, 160,163
Iron:
    advantages vs. disadvantages of,
        24, 26
    bell-and-spigot,
        cutting, 194
        described, 86
    cutting, 86, 90, 194
    as DWV piping material,
        21, 24, 26
    fittings for, 86
    hanging,
        of horizontal piping, 34,
        116-118
        of vertical piping, 116

J

Jigsaws, plumbing uses for, 10
Jim cap, as test fitting, 130

K

Kelly fittings:
    described, 21
    location of, 127-128

L

Layout, tools for, 4
Leaks:
    causes of, 75
    in gas lines,
        locating, 174-175
        repairing, 173-174
    in water-supply lines,
        locating, 75
        repairing, 75-77
Levels, plumbing uses for, 4

M

Manifolds:
    air-chamber, 143
    water-supply, on slab, 160-162
Mini-hacksaws, plumbing uses for,
    8
Miter saws, plumbing uses for, 59
Mule, for cutting and threading
    galvanized-steel pipe, 78, 80

N

Nail plates, as water-supply pipe
    protection, 141
National Energy Policy Act of
    1992, mentioned, 192
Nipples, right/left, 172, 173-174
Nut drivers, plumbing uses for, 3

EDITORS: JEFF BENEKE, RUTH DOBSEVAGE

DESIGNER/LAYOUT ARTIST: CATHERINE CASSIDY

ILLUSTRATOR: FRANK HABBAS

PHOTOGRAPHER, EXCEPT WHERE NOTED: JEFF BENEKE

TYPEFACE: ITC STONE SERIF